THE WASTE CRISIS

Landfills, Incinerators,
and the Search for a
Sustainable Future

Hans Tammemagi

New York   Oxford

Oxford University Press

1999

Oxford University Press

Oxford   New York
Athens   Auckland   Bangkok   Bogotá   Buenos Aires   Calcutta
Cape Town   Chennai   Dar es Salaam   Delhi   Florence   Hong Kong   Istanbul
Karachi   Kuala Lumpur   Madrid   Melbourne   Mexico City   Mumbai
Nairobi   Paris   São Paulo   Singapore   Taipei   Tokyo   Toronto   Warsaw

and associated companies in
Berlin   Ibadan

Published by Oxford University Press, Inc.
198 Madison Avenue, New York, New York 10016

Oxford is a registered trademark of Oxford University Press

Library of Congress Cataloging-in-Publication Data
Tammemagi, H. Y.
The waste crisis : Landfills, incinerators, and the search
for a sustainable future / by Hans Tammemagi.
p.   cm.
Includes bibliographical references and index.
ISBN 0-19-512898-2
1. Integrated solid waste management.   I. Title.
TD794.2.T36   1999
363. 72'85—dc21   98-53199

9 8 7 6 5 4 3 2 1

Printed in the United States of America
on acid-free paper

Today the environment is under seige. There is serious concern whether our children and grandchildren will be able to enjoy the delicate beauty of a blossom in spring, the call of a bird winging high on a breeze, or the graceful leap of a startled deer.

This book is dedicated to those individuals who not only want to understand waste management principles but will challenge them, break away from the common mindset, and develop new and improved ideas—ones that will benefit and sustain our environment for those that follow.

PREFACE

A long time ago I lived in a small northern town that was surrounded by primeval forest, cliffs of ancient Precambrian rock, and a maze of deep-blue lakes and creeks. The vast wilderness had a rugged beauty that brought a soul-fulfilling tranquillity to our lives.

When visitors arrived, they invariably wanted to visit the town dump to see the bears that were always present, snuffling and rooting through the garbage. But I was intrigued by the dump itself.

That tangled mound of refuse contained every conceivable relic of human usage, as though a tornado had torn through town, uprooted everything in its path, jumbled it all up, crushed it, and then dropped the whole lot in this spot. Pieces of furniture, threadbare tires, rusty appliances, amputated branches from yard maintenance projects, and ragged clothing formed a microcosm of our society. And scattered throughout the dump, all sorts of food wastes tumbled from torn plastic garbage bags making tasty morsels for the bears.

In a small town with no movies or bowling alleys, the dump served as an unofficial social center. Neighbors chatted with each other as they unloaded bulky items from trailers or car trunks. Kids rummaged through the refuse seeking treasure. And many came to enjoy the bears. The dump also competed with the local garage as the site for changing oil, with local do-it-yourselfers parked on slight inclines emptying old crankcase oil onto the ground.

The dump was usually veiled by smoke from fires that smouldered in one area or another. In the fall, temperature inversions often held the smoke close to the ground like a blanket. The eerie smoke and pungent smell drifted far out along the highway, announcing the presence of our town long before you reached it.

During my teenage years my pals and I would sneak into the dump after dark where, armed with powerful flashlights and 22-calibre rifles, we hunted rats. Sometimes we threw handfuls of bullets into the burning parts of the dump and scrambled to hide behind cars or trees where we expectantly awaited the ensuing explosions. We also experimented with batteries, gasoline, and other interesting fluids to create a sizzle, bang, or pyrotechnical display.

After a heavy rain, a murky liquid seeped from the dump into a nearby stream that meandered about half a kilometer through a stately pine forest until it reached the river. No one was concerned since the cloudy seepage usually cleared up in a few days. Occasionally, the health department placed restrictions on the amount of fish that could be eaten, but we were sure any problems were caused by industries up the river and not by our little dump.

I visited the old town not long ago and couldn't resist a drive out to the dump. A chain-link fence guards the perimeter now—I guess it's as much to keep humans out as it is to keep the bears away—and fires aren't allowed. But I heard that the kids still sneak in and occasionally the echo of gunshots breaks the solitude and leaping flames light up the vast night sky.

After we moved south, renovations on our new home soon created the need to visit the municipal landfill. Spread over a hundred acres, with trucks and bulldozers scurrying like ants over its vast girth, it made our northern dump look insignificant. Built in an abandoned quarry on top of a large cliff, it overlooked a neat middle-class suburb. Decades of blasting had weakened the underlying rock, creating fractures along which seepage from the landfill could flow.

Unlike our isolated northern community, the environmental problems caused by this much larger landfill were quite conspicuous. The local newspaper broadcast a continuing litany of sewer backups, creek contamination, and foul odors suffered by the residents living below the landfill.

This is not an isolated situation. Similar scenes are played out across North America since every village, town, and city must have at least one operating landfill, and there are many that are now closed. And what gargantuan size are the dumps for cities like New York, Toronto, Chicago, and Los Angeles?

In the last century the activities of humans have placed enormous stress on the environment. The exponential growth in population, combined with an increasing appetite for consumer goods, has led to an explosion in the amount of garbage we produce. At the same time, finding space for landfills is becoming more difficult.

Finding solutions to the garbage crisis is not simple. The problem is enormous in size, vital in terms of its impact on the environment, and complex in scope, involving not just many technical disciplines, but also social, political, and policy issues.

It is hard to believe that an everyday item such as garbage can be the source of so much misinformation, emotional debate, and misdirected effort. Sensation-seeking media and self-serving "environmental" groups have made it difficult to separate fact from fiction and have hindered the development of solutions to what is a growing, pervasive, and very real problem.

The purpose of this book is to provide a comprehensive overview of solid waste management in North America and to seek solutions to the waste crisis. The magnitude and complexity of the problem are explained, including a description of the quantities and character of solid waste being produced by society. The focus is on municipal wastes, but this is placed in the perspective of hazardous, biomedical, and radioactive wastes as well. The main components of an integrated waste management program are described, including recycling, composting, landfills, and waste incinerators. The scientific and engineering principles underlying these technologies as well as illustrative case histories are presented.

Unlike other textbooks, *The Waste Crisis* places this important subject into the broader societal context and discusses policies and strategies involved in the solid waste management field. It starts by questioning whether near-surface landfills are the appropriate solution for waste disposal, and then tries to find a path that is based on fundamental principles that can be applied to all waste types. It presents a framework based on sustainable development for making decisions about waste issues. The intent is to encourage the reader to constantly challenge commonly held perceptions and seek new and better ways of doing old things.

The Waste Crisis has been designed to be a general reference that is suitable for a wide readership. Because of its relatively detailed technical content, it can be used at the university and college level, particularly for students in the arts and science streams who are interested in learning about this topic but who will not be involved in the detailed design of waste systems. *The Waste Crisis* will also be a useful reference for engineering students who will benefit from the policy and social/environmental issues that are discussed.

Because *The Waste Crisis* is written in a non-mathematical style with numerous case histories, sidebars, and a comprehensive glossary, it will also be of interest to anyone who wishes to learn about this important topic or who has an interest in the environment. Thus, *The Waste Crisis* is appropriate for the bookcases of public libraries and high schools. The general reader who wishes to avoid technical details could skip chapters 6 (waste descriptions) and 10 (waste containment and treatment). Chapters 7 (landfills) and 9 (incinerators) are recommended reading, but could be skimmed rather than read in detail.

The Systeme Internationale (SI) units are used throughout the book. Because the British system of units is so firmly entrenched in the United States, the British equivalents are also presented in some cases (in brackets after the SI values).

I am grateful to the following for their assistance and support in making this book become reality. Early drafts were improved by the reviews of Dr. Vera Lafferty, Gunther Funk, Ian Brindle, John Orser, Frank Fohr, Dr. Francine McCarthy, and several others whose identities were never revealed. Joyce Berry and MaryBeth Branigan at Oxford University Press patiently and knowledgeably guided the editing process. I am indebted to my wife Allyson for her support throughout this long project.

The following organizations are thanked for providing information about their facilities or equipment and/or giving permission to use their photographs in this book: Brian Kearney Inc., Salt Lake City, Utah; R. Cave and Associates, Ontario; Chem-Security (Alberta) Ltd., Calgary, Alberta; ECDC Environmental, Salt Lake City, Utah; ECO Waste Solutions, Burlington Ontario; Guelph Engineering Dept., City of Guelph, Ontario; Lancaster County Solid Waste Management Authority, Lancaster, Pennsylvania; Midwest Bio-Systems, Tampico, Illinois; Mine Reclamation Corporation, Palm Desert, California; Municipal Archives of the City of New York; New York City Department of Sanitation, New York; Ogden Martin Systems of Lancaster Inc., Pennsylvania; Regional Municipality of Ottawa-Carlton, Ontario; Solmax Geosynthetics, Etobicoke, Ontario; and Wright Environmental Management Inc., Richmond Hill, Ontario.

Credit for photographs is given to the appropriate organization. Copyright remains with the organization credited for the photograph. Where no credit is given, the photograph was taken by the author.

CONTENTS

THE WASTE CRISIS

1

WASTE

We are a wasteful lot on planet Earth. We do not mean to be, but this is an inherent and unavoidable feature of human society. The processes of living, eating, working, playing, and dying all utilize consumer products whose production and use generate wastes. Every candy bar has a wrapper; every apple has a core.

It is almost impossible to think of a process that does not create some waste. There is sawdust from cutting lumber, metal shavings from drilling and soldering circuit boards, sludges from chemical processes, leftover food from restaurants, waste paper by the ton from environmental hearings and other legal proceedings, dirty diapers, and other household garbage. Societal wastes range from the refuse produced by every family to highly toxic industrial wastes from the production of specialized goods such as electronics, computers, cars, petrochemicals, and plastics. Virtually every aspect of our daily lives generates waste. Waste cannot be avoided.

But what happens to all this waste?

Some of it is recycled. During the past ten years there has been a growing realization that our globe is finite in its resources, and that the environment is under considerable stress and is being quietly but relentlessly despoiled. In response, streetside "blue box" and other recycling programs have sprouted (see Figure 1.1). Approximately 20% of municipal waste in North America is currently being recycled: metal cans are going back to smelters, paper back to pulp mills, and glass and plastic to factories to be turned into new products. Recycling programs are still expanding, and it is anticipated that in the future as much as 50%, and perhaps even more, of all household and commercial waste will be recycled.

3

1.1 Weekly garbage and recyclables at the curbside.

Some of the waste is incinerated. When this is accompanied by generation of electricity or useful steam or heat, it can be viewed as a form of recycling—the conversion of waste to energy, a very useful product. It also helps preserve precious nonrenewable resources such as gas, oil, and coal. Many people, however, are concerned about the emissions that are released into the air and the ash that is produced. About 18% of municipal solid waste in the United States is currently being incinerated, with about 75% of the incinerators generating energy (EPA, 1994). In Japan, approximately 34% of municipal solid waste is incinerated (Hershkowitz & Salerni, 1987); in Canada, the amount is less than 5%.

What happens to the waste that remains? As has been the practice for the past three millennia, almost all of it winds up in a landfill. Tens of thousands of landfills are dotted throughout North America, the final resting spots for the wastes we create in our everyday lives. Although efforts are now being made to reduce the amounts of waste by recycling and incineration, landfills continue to be the cornerstone of waste management. Figure 1.2 illustrates the three main methods for dealing with waste.

We have been reliant on landfills for so long that we automatically accept that they are the proper way of dealing with wastes. But is this really the case? Are landfills really serving our society and the environment well? A closer inspection shows that landfills are not the perfect solution that has been assumed.

Magnitude of the Problem

The waste disposal problem is closely related to population growth and urbanization. If we still lived in a world where every family had access to large tracts of land, there would be no difficulty in using a small fraction of the land for garbage disposal. But in this age of urbanization the situation is quite different, with hundreds of families often crammed onto a single acre.

A significant problem with landfills is simply their large numbers and the expanses of valuable real estate they occupy. Every city and town has one or more operating or closed disposal sites, although in recent years there has been a trend to fewer, but much larger, regional landfills. Industries and commercial operations have also used landfills as integral parts of their operations.

New York City's only landfill, Fresh Kills, is the largest in the world, covering an astonishing 1,000 hectares (2,500 acres) and reaching a height of 150 meters. Fresh Kills is built on a tidal flat, with no underlying liner to prevent leachate from seeping into the marine ecosystems. Aside from the damage to the environment, it is not clear what New York will do once Fresh Kills has reached full capacity.

Similar problems with waste disposal affect virtually all major centers, including Paris, Tokyo, and Mexico City. Near Marseille in France, Europe's largest waste dump spreads across 84 hectares (210 acres) and has grown to a mountainous height.

1.2 Recycling, landfills, and incinerators: the main components of waste management.

The total combined size of these landfills is mind-boggling. For example, in 1991 the province of Ontario (population approx. 11 million) had approximately 3,690 municipal and industrial landfills (MOE, 1991a); 37% (1,360) were active operating landfills, and 63% (2,330) were closed landfills. In addition, there are many unrecorded landfills that are now abandoned and forgotten. It is estimated that Ontario's landfills occupy a volume of approximately 1.1 billion cubic metres and a surface area of about 29,600 hectares (74,000 acres). This area could hold more than 190,000 single-family homes or produce approximately 2,600,000 bushels of wheat each year.

Even countries with seemingly unlimited wide-open spaces cannot afford to waste land on landfills. Although Canada occupies the second largest land mass of the countries of the world, prime agricultural land constitutes only about 0.8% of the total (about 72,090 square kilometers, slightly larger than the state of West Virginia). Furthermore, 43% of the value of agricultural production in Canada comes from farmland within 80 kilometers of 22 major metropolitan centers. In other words, good agricultural land is scarce, and much of it is close to major cities, placing it in direct competition with landfills, which are also close to the municipalities they serve.

Land as a Dwindling Resource

One way of visualizing the importance of land as a resource is to consider the amount of land that is available to each person on the globe. This is done by dividing the land area of the Earth (approximately 133 million square kilometers) by the total world population. Table 1.1 shows how the amount of land per person has changed since 1650.

We can see that by the year 2020, only 1.5 hectares will be available to provide space for housing, food production, waste disposal, and other needs for each individual. And since much of the Earth's land surface is occupied by mountains, deserts, Antarctica, and other unusable terrain, the amount of land per person available for beneficial activities is actually much smaller. Land is a diminishing resource that must be carefully protected.

Because decomposition generates explosive methane gas, and waste settles over time, it will not be possible to build any structures over landfills once they are closed. This rules out urban, commercial, or industrial development. Contaminants inside the land-

Table 1.1   Amount of land per person, 1650–2020.

Year	Land Area per Person Hectares (Acres)	
1650	24	(60)
1800	15	(38)
1900	8.3	(21)
1950	5.3	(13)
1990	2.4	(6)
2020	1.5	(3)

fills also preclude the use of these areas for agriculture. The vast areas covered by North America's closed landfills will be unsuitable for anything but recreation. This land represents a very significant resource that will not be available to us, to our children, or to many generations that follow. The situation is exacerbated since almost all these landfills are situated near or inside urban centers, where development pressure is greatest.

This unproductive use of land by surface landfills is contrary to the principle of sustainable development. New and different waste disposal approaches should be sought that conserve valuable land resources.

Life Inside a Landfill

What happens inside these large piles of refuse called landfills? Although out of sight, the waste is not as dormant as it appears to the casual eye. Instead, a landfill has a life of its own—the pulse is slow, but it ticks quietly and resolutely. Deep inside, microbes are feeding on organic materials and producing chemical changes. Settlement takes place as the lower parts of the landfill are compressed by the weight from above. The landfill settles, festers, and slowly decomposes.

Infiltrating rainwater leaches heavy metals from discarded batteries and other refuse to form a contaminated liquid that sooner or later percolates downward and pollutes the underlying groundwater. The U.S. Environmental Protection Agency estimated that of the 75,000 landfills in the United States, 75% are leaking (Lee & Jones, 1991). Damage to the groundwater is often very serious—and usually irreversible—because landfills contain materials that are far from benign.

The variety of wastes, particularly those that are harmful, is staggering. They include chemicals such as PCBs, lead, solvents, dioxins, DDT, benzene, CFCs, furans, and many more. Today around 45,000 different chemicals are produced by industry, and about 1,000 new ones are added each year. Unfortunately, many of these chemicals are toxic: that is, they are poisonous to humans and can also damage other parts of the complex and fragile environmental web.

Landfills also cause atmospheric pollution. The decomposition of waste creates gases such as methane, carbon dioxide, vinyl chloride, and hydrogen sulphide, which slowly seep into the air around the landfill. This impairs air quality in the immediate vicinity and, on a larger scale, contributes to the greenhouse effect and global warming.

Even modern landfills that employ state-of-the-art technologies such as liners and leachate collection systems are a problem. If they are not leaking now, they will probably start leaking within a few decades of their closure. The use of modern technologies simply postpones the inevitable.

Looking for Solutions

In the past, convenience was the motivating factor in landfill siting and design. The philosophy behind landfills was "out of sight and out of mind." This approach has been used for centuries, and it may have been acceptable when the Earth seemed infinitely large and contained few people. Now, however, it is being increasingly recognized that near-surface landfills, the cornerstone of almost all waste management systems, are not really out of sight and out of mind; instead, they are inadequate and are placing a long-term burden on the environment.

But how do we avert the waste crisis that is quietly approaching? How do we break out of patterns that have been so firmly entrenched?

This book does not have a magical solution, but it does provide background and technical information so that the reader can understand the scope of the problem. More important, this book explores a number of alternatives and tries to stimulate the reader to question the way things are currently being done and to search for new and innovative solutions. We need to think "outside the box."

Discussion Topics and Assignments

1. Arrange for a tour of your local landfill. What environmental impacts can you observe? What other environmental impacts are you aware of?
2. It is estimated that by 2025 there will be 93 cities in the world with populations exceeding 5 million. How will these cities manage their solid waste?
3. What evidence is there that the human population has reached or surpassed the size where the world can accommodate it?
4. Are you interested in a career in solid waste management? Why? Identify the federal, state/provincial, and local agencies that are responsible for solid waste management in your area. Which would you prefer to work for, and why?

Suggested Reading

Carson, Rachel. 1962. *Silent Spring.* Cambridge, Mass.: Riverside Press.
Goldsmith, E. A. R., Allaby, M., Davol, J., and S. Lawrence. 1972. Blueprint for survival. *The Ecologist, 2,* 1–43.
Horton, Tom. 1995. Chesapeake Bay: Hanging in the balance, *National Geographic,* June, 2–35.
Lee, G. F., and R. A. Jones. 1991. Landfills and ground-water quality. *Groundwater, 29,* 482–486.
Meadows, D. H., Meadows, D. L., Randers, J., and W. W. III Behrens. 1972. *The Limits to Growth (A Report to the Club of Rome).* New York: Universe Books.
Ward, B., and R. Dubos. 1972. *Only One Earth: The Care and Maintenance of a Small Planet.* New York: Norton.
Whitaker, J. S. 1994. *Salvaging the Land of Plenty: Garbage and the American Dream.* New York: William Morrow.

2

STARTING FROM BASICS

Sustainable Development

The landfill has been a child of convenience. Historically, waste was simply dumped in depressions, ravines, and other handy locales that were close to the population centers producing the waste. For centuries this was an acceptable method, but two developments caused serious environmental difficulties with this approach. First, the enormous growth in population resulted in much more garbage being generated, at the same time as land was becoming a scarcer and more valuable resource. Second, the technological and consumer revolution led to the creation of many more hazardous products—particularly synthetic organic substances such as pesticides, PCBs, paint removers, and degreasers, which ultimately wound up in landfills. Landfills grew bigger, and their contents were more toxic than ever before. The child of convenience grew up and turned into an environmental ghoul.

Instead of convenience, we need to seek methods of waste disposal that do not impair our environment, use up valuable resources, or place limitations on future resources. Changing engrained habits is not an easy task. We need a revolution that sweeps aside the old ways and introduces new concepts and technologies that are in accord with philosophies that value and protect our environment.

Although the gravity of the situation is becoming recognized, and some positive steps—such as streetside recycling programs—are being implemented, there is still an enormous amount to be done. Perhaps we need a different outlook on waste disposal. We should seek disposal technologies and methods that protect the environment; furthermore, these methods must be based on fundamental philosophies that the public understands, agrees with, and buys into.

When we seek to redesign waste management, it is important to start with the ultimate objectives firmly in mind. We need goals and a set of rudimentary principles to guide us.

Many of us have read a science fiction novel in which a lonely spaceship has been sent to explore a distant galaxy, hundreds of light years away in the farthest reaches of the known universe. Even at hyperspeeds, the spaceship must travel for centuries to reach its destination, requiring several generations of crew to pass their lives aboard the ship. To make this possible, the ship has been designed to be totally self-contained. Energy is obtained by solar cells. The spaceship recycles everything: no refuse is jettisoned; every molecule is recycled. The spaceship has attained the ultimate in self-reliance and sustainable development.

We need to set a similar goal of sustainability for spaceship Earth. In fact, if the human race is to continue to grow and wants to survive, this vision must become reality. We are currently a long way from achieving this goal.

Sustainable development is a concept that has received increasing acceptance in recent years and has become the cornerstone of many government policies. The United Nations endorses sustainable development as the philosophy that should guide nations in the conduct of their commercial and industrial activities. Since waste management activities are a subset of the overall industrial undertakings of society, it is important that they should also be governed by this philosophy.

The concept of sustainable development can be viewed as the integration of economic, social, and environmental systems. Conventional economics maximizes the goals of the economic and social systems, but sustainable development maximizes the goals of all three systems.

The United Nations' World Commission on Environment and Development defines sustainable development as follows (World Resources Institute, 1992): "Sustainable development is development that meets the needs of the present without compromising the ability of future generations to meet their own needs."

The sustainable development concept is generally applied to the "front end" of the industrial cycle—the aim of ensuring that humans can feed, clothe, house, and entertain themselves. It is particularly relevant to preserving our diminishing stock of natural resources such as forests, minerals, oil and gas, and fish stocks.

How Did Sustainable Development Evolve?

In the 1970s, governments began to recognize that environmental problems—such as the greenhouse effect, ozone depletion, groundwater depletion and pollution, deforestation, desertification, and species extinction—were beginning to threaten the globe. This concern was first articulated in the United Nations Conference on Human Environment held in Stockholm in 1972, which established the United Nations Environment Program, whose responsibility was to build environmental awareness and stewardship.

The independent World Commission on Environment and Development was also established with the mandate to look at how development affected the environment. The commission was chaired by Gro Harlem Brundtland, then minister of environment and later prime minister of Norway. Their report, *Our Common Future*, was issued in 1987 and contained the first use of the term "sustainable development." This concept soon captured the world's imagination because it promised a solution to the dilemma of global population growth and limited environmental resources.

In 1988, more than 50 world leaders supported the report, and in 1992, the first United Nations Conference on Environment and Development, also known as the Earth Summit, was held in Rio de Janeiro. The Earth Summit provided enormous publicity and support for sustainable development and many initiatives emerged to implement this concept. The initiatives led to ISO 14000, the international protocols for companies to develop and incorporate environmental management systems.

Although the United Nations World Commission emphasizes decreasing the pollution discharged by industrial plants, it does not mention improving waste disposal practices. This "blind spot" is typical of our society's "out of sight, out of mind" attitude toward landfills.

Sustainable development must also be applied to the "back end" of the consumer cycle: the management and disposal of wastes. For example, leaking landfills degrade surface waters and groundwater, some of the most basic and valuable natural resources we have. This is *not* sustainable development. The space occupied by thousands of landfills displaces millions of acres of land from other uses, such as agriculture and urban development. This is *not* sustainable development. Leaving a legacy of leaking landfills will require our grandchildren to utilize their intellectual effort, their time, and their resources to provide ongoing repairs and maintenance. This is *not* sustainable development. It is simply ignoring the problem and, through procrastination, deferring it to our children and their children.

It is important that we alter our approach to pursue a complete, holistic concept of sustainable development that includes not only the natural resources at the front end, but also the disposal of wastes at the back end. Sustainable development must form the cornerstone of our approach to waste disposal in the future.

General Principles

We need to develop disposal systems that do not leave a legacy for which our grandchildren must accept responsibility. Just as mathematicians construct a mathematical proof, we need to start with a fundamental truth—a fundamental principle—and logically and systematically build on it to emerge with a detailed action plan.

Based on the underlying principle of sustainable development, three additional, more specific principles can be deduced for waste management. Although these general principles appear simple and almost self-evident, they provide powerful guidance to the practical design and development of waste management systems. They are described below, with some preliminary discussion of how they affect waste management and disposal.

Protect Health and Environment
Waste management and disposal must be conducted in a manner that does not pose a risk to human health or the environment, either now or in the future.

This principle places important constraints on the siting and design of disposal facilities and also on the form of the waste. For example, it effectively requires that leakage of leachate into groundwater and emission of landfill gases into the atmosphere should be eliminated, or at least reduced to an amount that the environment can assimilate safely. Since the large majority of landfills either are or will soon be leaking leachate and/or emitting gases in an uncontrolled fashion, the implementation of this principle will not be simple. We will need to look at dramatically different ways of disposing of our wastes.

Minimize the Burden on Future Generations
Wastes should be managed in a way that does not place a burden on future generations.

This principle protects future resources. That is, our grandchildren should not have to spend their time and effort looking after the wastes bequeathed by our generation. This principle is a paraphrase of the U.N.'s definition of sustainable development.

Conserve Resources
Nonrenewable resources should be conserved to the maximum extent possible.

There are two ways in which this principle applies to waste management. First, the process of managing and disposing of wastes should not consume nonrenewable resources. In particular, this principle recognizes that land is a valuable natural resource that must be protected.

Second, this principle requires that all useful resources should be extracted from waste prior to disposal, to the extent practical. The extraction of recyclable materials from waste also minimizes the amount of waste that requires disposal, which has numerous benefits, as we will see. This principle is another way of stating that recycling is an important and fundamental part of waste management.

These three principles have been derived from the fundamental goal of integrating waste management with the philosophy of sustainable development. The relationship is illustrated conceptually in Figure 2.1. In turn, more detailed guidelines can be derived from these principles. The combined principles and guidelines will provide a logical basis for planning and implementing waste management systems, in contrast to the current de facto principle of convenience.

We need a new approach to waste disposal that is based on the following fundamental principles:

- Protect health and environment
- Minimize the burden on future generations
- Conserve resources

Public Involvement

Who is to blame for the looming waste crisis? The engineers and scientists who design and build disposal facilities? The government officials who develop and enforce regulations? The politicians

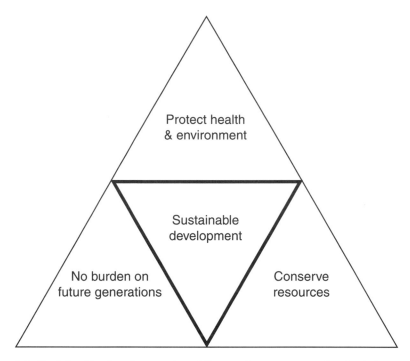

Protect health
& environment

Sustainable
development

No burden on
future generations

Conserve
resources

2.1 Sustainable development and the waste management principles.

whose jurisdictions own and operate these facilities? Or the public who generate the waste and for whose benefit the waste facilities are intended?

The answer is that we are all to blame. Wastes are very democratic—they are produced by each and every one of us, and so we should all contribute to the solution. Because of the populist nature of waste, its management is very much a social problem. It is not sufficient to understand the technical aspects; it is equally important to come to grips with the social and political issues. Interwoven with these are economic issues, and we must seek effective solutions within practical cost limitations.

Unlike other complex projects, such as the space shuttle, effective future waste management cannot be turned over to a highly skilled project team working with great focus and motivation but in isolation. Instead, the public should be intimately involved in the undertaking. Without their support, both directly and through their elected representatives, no progress will be made. This marriage of technical and sociopolitical disciplines brings challenges as well as fascination to the field of waste management.

Discussion Topics and Assignments

1. "Sustainable development" has become a hot buzzword. Keep track of how many times you see or hear it over a one-week period, and in what context.
2. Do you think that sustainable development is as important or fundamental as is generally thought? Why?
3. Identify some important benefits and drawbacks of living in an advanced industrial country.
4. Do you feel that the world's population growth should be reduced and the world's population stabilized as soon as possible? What is an appropriate world population? What policies should be instituted to accomplish this?
5. What are your responsibilities to future generations? List important environmental benefits and detriments that have been passed to this generation by predecessors. List important environmental benefits and detriments that this generation will pass to future generations.

Suggested Reading

Ehrlich, Paul R., and Anne H. Ehrlich. 1988. Population, plenty, and poverty. *National Geographic*, December, 914–945.

Government of Canada. 1990. *Canada's Green Plan for a Healthy Environment*. Ottawa: Minister of Supply and Services.

Miller, G. T., Jr. 1990. *Living in the Environment: An Introduction to Environmental Science*. Belmont, Calif.: Wadsworth Publishing Company.

Turner, R. K. (editor). 1988. *Sustainable Environmental Management: Principles and Practice*. London: Belhaven.

World Commission on Environment and Development. 1987. *Our Common Future* (*The Brundtland Report*). Oxford and New York: Oxford University Press.

3

HISTORICAL PERSPECTIVES
What Can We Learn?

More than any other single event, the seemingly endless wandering of the garbage barge *Mobro 4000* symbolizes the frustrating situation we find ourselves in. The barge, laden with refuse from the town of Islip on Long Island, New York, set sail on March 22, 1987, and roamed for 55 days from port to port down the Atlantic seaboard, along the coast of Central America, and into the Caribbean in search of a place that would accept its smelly load. None would. Eventually, after having traveled more than 9,600 kilometers, the barge returned to New York, where the waste was finally incinerated and the ashes placed in a landfill.

A garbage crisis is at hand. The situation has not improved since the *Mobro* incident. As a society we are generating far too much waste, especially in North America. At the same time, places to dispose of it are becoming limited. The public and politicians have recognized the inherent dangers of existing landfills and are refusing to build new ones—or, as in the case of the *Mobro*, they are refusing to accept any more waste than is necessary. How did we get into such a mess?

Garbage through the Ages

The first recorded regulations to control municipal waste were implemented during the Minoan civilization, which flourished in Crete from 3000 to 1000 B.C. Solid wastes from the capital, Knossos,

19

were placed in large pits and covered with layers of earth at intervals (Wilson, 1977). This basic method of landfilling has remained relatively unchanged right up to the present day.

In Athens, by 500 B.C. it was required that garbage be disposed of at least 1.5 kilometers from the city walls. Each household was responsible for collecting its own garbage and taking it to the disposal site.

The first garbage collection service was established during the period of the Roman Empire. Householders tossed their refuse into the streets, and then it was shoveled onto horse-drawn carts and transported to an open pit, often located within the community. The bodies of dead animals (and sometimes people) were buried in pits outside the towns to spare inhabitants their odor.

After the decline of the Roman Empire, there was no organized method of waste disposal in Europe for several centuries. Through the Dark Ages and the Renaissance, as long as land was plentiful and people were few, garbage was simply dumped in convenient, out-of-the-way places and forgotten. The attitude through much of history was that nature could take care of itself. This philosophy manifests itself today in sayings such as "Dilution is the solution to pollution," as well as in waste dumping into the oceans, which are seen as vast, unspoilable resources.

As Europe's population grew and became more urbanized, land and natural resources became scarcer, and the impact of garbage became more significant. Sweepers were hired to rake refuse and dung from city streets and load it into carts which carried it away from the city. Another convenient way of removing garbage was to throw it into waterways and rivers. This created many problems and undoubtedly contributed to epidemics such as the Black Death, which struck England in 1347. In 1388, the English Parliament banned the disposal of wastes into public waterways and ditches, although the practice continued (illegally) well into the 19th century.

Europe, with its concentrated population and large cities, began to experience a garbage crisis around 1500. At that time, garbage and offal from households, manure from stables, and refuse from industry were simply dumped into the central gutter of the street. Ravens, dogs, and pigs roamed the streets and lived well by scavenging. It was reported that the garbage piled up so deep outside the gates of Paris that it began to interfere with the city's defenses.

The Black Death

The plague that took a greater toll of life than any of Europe's wars or natural catastrophes was called, appropriately, the Black Death. Ravaging Europe between 1347 and 1351, it killed as much as one-quarter of the population of England in the initial outbreak. It is estimated that the plague may have killed 25 million people out of a European population of 80 million. Originating in China, the plague was transmitted to Europe when a Chinese army unit catapulted disease-infested corpses into an outpost that it was besieging in Crimea. The plague soon spread throughout Europe, exacerbated by the poor sanitation of the time. There were recurrences of the plague every few years until 1400. With its immense death toll, the Black Death affected all aspects of life and caused significant social upheaval.

By the 1700s, refuse had become a major problem and complaints were numerous. In 1741, for example, Lord Tyrconnel described the streets of London as "abounding with such heaps of filth as a savage would look on with amazement." In 1832, outraged citizens complained about streets in the vicinity of Westminster Abbey which were "the receptacle of all sorts of rubbish which lay rotting and corrupting, contaminating the air and affording a repast to a herd of swine."

In the 19th century, wastes were still dealt with in the most rudimentary fashion: dumping in the streets or other convenient places was the order of the day. Mounds of garbage rotted in the streets and alleyways. Open burning of garbage by individual homeowners and companies was a common practice. Increasingly, public health officials linked sanitation practices such as inadequate garbage disposal to the incidence of disease and other health risks in urban areas. A report published in England in 1842 linked disease to filthy environmental conditions and helped launch the "age of sanitation."

Support grew for municipalities to take control of urban wastes, which had formerly been viewed primarily as an individual responsibility. In the United States, the modern concept of solid waste management first emerged in the 1890s in response to the sanitation problems associated with rapid industrialization and urbanization in the second half of the nineteenth century (Blumberg & Gottlieb, 1989).

America After the Turn of the Century

In North America, with a much smaller population density and less industrialization, garbage problems were slower to evolve than in Europe. Nonetheless, the need for proper sanitation and garbage management was highlighted by an outbreak of yellow fever in Memphis, Tennessee, in 1878 that claimed more than 5,000 people out of a population of about 40,000. In response, Colonel George Waring, Jr., installed a system of municipal sewers which had many unique features, including daily flushing (Malone, 1936).

By the turn of the 20th century, a growing number of U.S. cities provided at least a rudimentary level of solid waste collection and disposal. Colonel Waring moved to New York City, where he became commissioner of the Department of Street Cleaning and implemented some far-reaching reforms that greatly improved the cleanliness of the city's streets. The rates of death and disease declined substantially, particularly the incidence of diarrheal diseases.

George E. Waring, Jr.—The First Garbage Hero

Born in 1833 in New York, George Waring studied agricultural chemistry. He fought in the Civil War, gaining the rank of colonel. He is credited with installing sewer systems in Memphis, Tennessee, and, while street cleaning commissioner for New York City, he introduced three-part separation and collection of garbage. Waring can be regarded as the first hero in a business that is not easily romanticized. It is ironic that he died in 1898 from yellow fever, which he caught while studying ways to improve sanitation in Cuba.

A 1902 survey of 161 U.S. cities by the Massachusetts Institute of Technology showed that 79% provided regular collection of refuse. By 1915, 89% of major American cities had some kind of garbage collection system, and by 1930 virtually all large cities offered garbage collection services.

Once removed from urban centres, the wastes were disposed of in a variety of ways, ranging from landfilling (the most popular method) and water and ocean disposal to incineration. In some small to medium-sized towns, piggeries were developed where swine were fed fresh or cooked garbage.

There was a growing awareness that many of these waste management practices were inadequate. Water and ocean dumping came under the strongest attack, and by 1933 ocean dumping had been ruled illegal for municipal solid waste, although industrial and commercial wastes were exempted. From the 1880s to the 1930s, land dumping was the most prevalent method of waste disposal, although concerns were already being raised about the health risks associated with large open dumps. In the 1920s it was common to landfill by reclaiming wetlands near cities, using layers of garbage, ash, and dirt. It is interesting to note, though, that waste reduction through source separation and recycling was very much in vogue in the early 1900s.

An innovative method of waste management, introduced near the turn of the 20th century, was reduction. This involved cooking garbage to extract a variety of marketable byproducts such as grease and "tankage," a dried animal solid sold as fertilizer. However, many of the plants did not attain their expectations, and there were odor problems. Few reduction plants remained in operation after World War I.

The British and the Germans led the way in developing the practice of incineration, both to reduce the volume of waste and to produce energy. The first systematic incineration of municipal garbage took place over a century ago in Nottingham, England in 1874 (Murphy, 1993). The first plant to generate electricity from incineration was developed in Great Britain in the mid-1890s. By 1912, approximately 76 incinerators were generating electricity in Britain and another 17 were operating on the Continent.

Although a pilot incineration project was constructed in New York City in 1905, it was not until the decade after 1910 that incineration came into widespread use in the United States. At its peak in the 1930s, between 600 and 700 U.S. cities had constructed incinerators, and this became a significant method of disposal of municipal waste.

Although alternative methods were tried, landfills remained the most common form of waste disposal owing to the low cost of land and the simple, inexpensive technology involved. Landfills were generally established in shallow depressions, wetlands, and tidal flats. In the early to mid-1900s there was not much technology involved. Underlying liners were almost never used, and the landfills were usually only a meter or so thick. (Figure 3.1 shows landfills operating in the early 1900s.) As landfills expanded, growth took place horizontally and consumed large tracts of land. This growth, along with the inherent dirtiness of landfills, inexorably led to a conflict with urban development.

3.1   Historical photos showing landfills in early 1900 (Provided by Wright Environmental Management Inc., original source unknown).

The Age of Consumerism and the Waste Explosion

Two developments in the post–World War II era led to significant escalation in the problems of managing waste. First, a new phenomenon called "consumerism" emerged. A long period of prosperity, combined with improvements in manufacturing methods—particularly in mass production—led to rapid growth in the number and variety of consumer goods. In addition, new marketing and production practices were introduced, such as planned obsolescence and "throw-away" products. The growth of advertising, along with the electronic media, played an important role in the evolution to our society's current level of overconsumption. The end result was a dramatic increase in the amount and variety of consumer goods—and, hence, garbage. To compound the problem, packaging became a dominant force in the way goods were marketed, distributed, and sold. Today, packaging represents more than one-third of the entire waste stream (Blumberg & Gottlieb, 1989).

The second development was the birth of the "chemical age," which resulted in a dramatic change in the composition of the waste stream. The Japanese capture of the world's supply of rubber during World War II forced petrochemical companies to invent a wide range of replacement products that could be manufactured from crude oil. The petrochemical industry has grown explosively since that time, yielding a vast array of new synthetic organic compounds, including nylon, rayon, polystyrene, polyethylene, chlorinated organics such as polychlorinated biphenyls (PCBs), insecticides, and a host of other toxic chemicals. A kind of pollution that had never existed before entered the environment, and nature and humans are having difficulties adapting to it (see, for example, Carson, 1962).

Although the age of consumerism led to a huge improvement in lifestyle, it was accompanied by a dramatic increase not only in the quantity of wastes generated, but also in their toxicity.

The Chemical Industry

The chemical industry touches all aspects of modern society. Chemicals are used in the production of almost all consumer goods, including medications, cosmetics, appliances, fuels, plastics, electronic equipment, pesticides, textiles, and much more. The raw material for most of these products is fossil fuel, primarily petroleum and natural gas, so the chemical industry is also

→

The Chemical Industry (continued)

called the "petrochemical" industry. Petroleum and natural gas are turned into intermediate-process chemicals, which in turn undergo a variety of reactions to produce new chemicals—end products such as paints or cleaning fluids, or materials to manufacture other goods, such as plastics for computer keyboards and telephones.

Before World War II, there was virtually no manufacture of synthetic chemicals, but the petrochemical industry has grown vigorously since then. In 1950, about 1.8 million metric tons of basic chemicals were produced; by 1986, annual production had risen to nearly 90 million metric tons. The variety of new organic and hydrocarbon chemicals, such as PCBs and pesticides, also increased enormously over this period, accompanied by heavy pollution of the environment.

The globe and its ecosystems, which usually absorb changes over centuries and millennia, have never experienced such a sudden infusion of new chemicals. It is not surprising that fragile ecosystems are under great stress.

Until the mid-1900s, landfills were little more than open pits which were breeding grounds for rats and other vermin. Open burning was a common practice. These landfills were a public nuisance as well as a health hazard because of the vermin, windblown litter, odors, and out-of-control fires. Many landfills were near growing urban areas and their water supplies. Not unexpectedly, public opposition to these landfills began to grow.

To counter the mounting criticism, the concept of the "sanitary landfill" was introduced in the 1950s. A sanitary landfill is usually defined as an engineered method of disposing of solid wastes on land by spreading the waste in thin layers, compacting it to the smallest practical volume, and covering it with soil at the end of each working day (Stone, 1977). The simple step of covering the waste with a thin layer of earth at regular intervals was particularly important: it alleviated problems such as uncontrolled fires, windblown refuse, and rodents.

Another important feature was the requirement for an "engineering method." This implied that scientific and engineering principles were to be used, rather than simply dumping garbage in a convenient location. Data on waste quantities and generation rates, available cover soil, land use, transportation routes, and so forth were to be used to assess the suitability of sites for a landfill. Furthermore, the design and construction of facilities were to take advantage of natural conditions to protect the environment.

Although it was a positive step, the sanitary landfill still had some basic shortcomings. It did not satisfactorily address groundwater contamination, gas emissions, and related health concerns. Furthermore, many communities continued open burning and open dumping. According to the U.S. federal Bureau of Solid Waste Management, 94% of all land disposal operations in the mid-1960s were inadequate in terms of air and water pollution, insect and rodent problems, and physical appearance.

By the 1970s and 1980s, there was growing recognition that landfills were causing significant contamination of groundwater. Groundwater, the supply tapped by wells, is a vitally important resource which, when used as a drinking or irrigation supply, can directly impact human health. The problem was compounded by the fact that once groundwater becomes contaminated, it is exceedingly difficult to remediate.

Seeking solutions to the groundwater problem, landfill managers turned to technology and engineering. Although the basic landfill remained essentially unchanged, a number of features were added. Bottom liners made of clays or synthetic materials such as impermeable high-density polyethylene were introduced to stop leachate from leaving the landfill. Caps made of similar materials were placed over the landfill to decrease the infiltration of precipitation. In addition, engineered collection systems were installed to capture leachate and gas. Monitoring of groundwater, surface water, and gas emissions became a routine part of landfill operation.

In spite of these technical advances, there was continuing concern about groundwater contamination. Several studies during the late 1970s pointed out that leaking leachate was a problem facing all landfills. The U.S. Environmental Protection Agency estimated that in 1990, there were about 75,000 landfills in the nation, and more than 75% of them were polluting groundwaters with leachate (Lee & Jones, 1991).

It was also recognized that even state-of-the-art municipal landfills with double liners and other modern leachate containment systems would fail eventually. The increased use of engineering techniques would only postpone, not prevent, the onset of groundwater contamination.

Awareness grew of the hazard that municipal landfills pose. The concentration and toxicity of the pollutants found at municipal sanitary landfills were seen as capable of causing as great a risk of cancer as those from industrial waste landfills (Brown & Donnelly, 1988). In addition to grass clippings and old newspapers, the sites contained toxic chemicals in the form of used motor oil, oil-based paints, batteries, cosmetics, solvents (including oven cleaners and fingernail-polish remover), pesticides, tires, and much more. These substances contain chlorinated organic compounds such as vinyl chloride and trichloroethylene, as well as heavy metals, copper, and lead. Vinyl chloride is a gas that can cause liver cancer and neurological disorders, and lead can damage the nervous and reproductive systems (Parmeggiani, 1983).

Incinerators had been in vogue during the 1930s. The U.S. Air Quality Act of 1967, however, introduced new air emission standards that forced operators of older incinerators to add air pollution devices such as scrubbers and precipitators. Since incinerators were already more expensive and capital-intensive than landfills, this act essentially priced incineration out of the market. Within five years of the act, 100 incinerators had been shut down, and nearly all the rest followed within the next few years.

The 1970s were a decade of transition in the solid waste arena in North America. Landfills were coming under increasing attack, and the promise of energy-from-waste incinerators had not been fulfilled. The consumerism lifestyle and a long period of affluence led to the production of unprecedented amounts of garbage. By the 1980s, a crisis atmosphere had developed. The public recognized that municipal wastes were not being managed adequately. Waste volumes were growing, landfills were polluting the groundwater, and incinerators were expensive and were polluting the air. Serious environmental incidents at Love Canal, New York, Times Beach, Missouri, and Glen Avon, California, although not involving municipal landfills, served to focus public and media attention on all environmental issues.

The Crisis

In 1976, the U.S. Congress passed the Resource Recovery and Conservation Act (RCRA). Although much of the act dealt with hazardous wastes, Subtitle D addressed municipal solid waste. It required the Environmental Protection Agency to set standards for landfills based on adverse affects on health or the environment, while the states were required to catalog their open dumps and either bring them up to standards or shut them down. A period of five years was allowed for shutting down noncompliant landfills and dumps. This timetable caused a crisis at local levels and led to protracted legal battles to find new sites or extend the capacity of existing ones.

By the 1980s, the problem of waste disposal was a national issue which permeated all levels of government. To deal with this crisis, a plethora of laws were passed, creating a complex regulatory environment in which federal, state, regional, and local agencies had the authority to review and authorize landfills and other waste management facilities.

The number of municipal landfills in the United States dropped dramatically, from about 20,000 in 1979 to about 5,300 in 1993 (Miller, 1997). The primary cause was the stringent new guidelines, reinforced by an inability to site new landfills in the face of growing public opposition to having landfills in their communities.

In this setting, it is understandable that incineration with production of energy would appear to be an attractive alternative. In the early to mid-1980s interest in incineration reawakened, and approximately 100 new plants were committed and another 200 planned in the United States.

Resource recovery, in the form of recycling and energy recovery through incineration, also became popular in the mid-1980s. Today, most North American communities operate streetside or "blue box" programs for fine paper, cans, plastics, newspapers, and cardboard. The specific collection of household hazardous wastes—either on periodic collection days or at special collection sites—has also become a standard part of municipal waste service.

The importance of selecting a site that minimizes the environmental impact of a landfill began to be recognized in the 1980s. New siting criteria emphasized the importance of sites that were well above the groundwater table, that were not in groundwater recharge zones, that had low hydraulic gradients, that were not in natural flood areas, and that had natural impermeable clay formations to prevent contaminant migration. Over the past 15 years, the siting of landfills has become a sophisticated process which incorporates technical as well as social and political concerns. Continuing improvement in the siting process has greatly restricted the locations where landfills can be placed.

The construction of new landfills was about to become even more difficult. As urban areas expanded rapidly, it was only a matter of time until serious opposition to landfills began to mount. Community and neighborhood groups emerged in the 1980s and began to organize with a very specific goal in mind—to stop the construction of landfills (or any other waste facility) in their areas. Their community protests were augmented by litigation and lobbying, which took advantage of the plethora of laws and regulations being issued. This was very much a grass-roots development.

Although centered in the neighborhoods that were directly involved, the networks and coalitions established included practitioners adept in the politics of democratic mobilization. They could respond effectively to requests and provide assistance from other areas. The opposition to landfills and related waste management facilities, such as incinerators, has been highly successful in either stopping projects altogether or delaying them for a long time.

It is now a fact of life that any new waste management project will be met by strong local opposition. This kind of opposition has been dubbed NIMBY ("Not In My Back Yard"). People seeking a career in solid waste management need to have a good understanding of this phenomenon and how to work with it. The NIMBY phenomenon is discussed further in chapter 12.

The resistance to landfills has been effective, and few new landfills have been constructed during the 1990s. This has caused existing landfills to grow larger—usually upward—as their ability to expand horizontally has been limited. The end result is often negative, because many of these older landfills, which have now become superdumps, do not have adequate liners or are not situated in places where the environment can be protected.

Summary of Landfill Evolution

Landfill design evolved as a series of responses to problems. Only when a problem was identified or reached a sufficient level of concern were corrective steps taken. These improvements were invariably driven by regulatory requirements.

The evolutionary process in landfill design has been relatively simple and has involved only three significant changes, which are summarized in Table 3.1.

The first was the implementation of daily cover in the "sanitary" landfill, a response to the problem of odors, windblown refuse, open and uncontrolled fires, and rodents and other animals. These problems are categorized as health/nuisance factors. The term "sanitary" was somewhat optimistic, given the problems that remained unresolved.

Table 3.1   Summary of landfill evolution

Date	Development	Problems	Improvements
1970s	Sanitary landfill	Health/nuisance, i.e., odor, fires, litter	Daily cover, better compaction, engineered approach
Mid 1980s	Engineered landfill	Groundwater contamination	Engineered liners, covers, leachate and gas collection systems
Late 1980s	Improved siting	Groundwater contamination	Incorporation of technical, socio-political factors into siting process
Future (?)		Air emissions, long-term safety	(?) (?)

The second major step was the development of engineered liners and covers, both synthetic and natural, as major components of a landfill, in response to concerns about groundwater contamination. Although a positive step, these are now seen as only short-term solutions.

The third major step was not an improvement in design, but rather a recognition that a landfill forms a system integrated with its surrounding environment. This recognition led to significant improvements in the siting process and the location of landfills.

It is interesting to speculate how landfills will evolve in the future. To do so, we must identify the problems that remain unresolved. One of these, air emissions, is just beginning to rise to the level of public concern. The U.S. Environmental Protection Agency (EPA) has recently formulated standards for the collection and control of landfill gases. With the global issues of ozone-layer depletion, global warming, and ground-level smog, there will undoubtedly be a spread of EPA standards to other countries and jurisdictions. We can predict that in response to these regulations, numerous new emission-control technologies will be developed.

Another area of future concern, and one that will be explored further in later chapters, is the integrity of landfills over the long term—several centuries. Landfills at the ground surface and near the major urban centers they service provide safety only for the short term, even when they are well sited and have the latest engineered improvements. They simply cannot withstand the long-term effects of natural erosion and the remorseless encroachment of urbanization.

In summary, we are caught in a two-pronged problem: we generate too much waste, and we dispose of it inadequately. It is time for a hard, critical look at how we approach this problem, with a recognition that we must take some drastic steps and use some different strategies.

Discussion Topics and Assignments

1. From historical records, trace the evolution of waste disposal/management in your community or region over the past 50 years (or longer, if possible). Identify the major changes that have occurred, and explain why.
2. Waste management technology has not kept pace with the advances seen in other sectors, such as transportation and communications. Why is this so? What can be done to rectify the situation?
3. Identify and discuss new issues that will become important in solid waste management in the next few decades.
4. How do you think that incinerator technology will evolve and improve in the future?
5. Brainstorm ways of improving the landfill to provide better long-term containment and safety.

Suggested Reading

Brimblecombe, P. 1987. *The Big Smoke: A History of Air Pollution in London since Medieval Times.* London: Methuen.

Murphy, Pamela. 1993. *The Garbage Primer.* New York: League of Women Voters, Lyons & Burford.

Wilson, D. G. 1977. The history of solid waste management. In *Handbook of Solid Waste Management*, edited by D. G. Wilson. New York: Van Nostrand Reinhold.

4

INTEGRATED WASTE MANAGEMENT
More than Just Landfills

What Is Integrated Waste Management?

Just as a general fights a battle with tanks, infantry, artillery, and air support, the campaign against waste also requires an arsenal of many weapons. Instead of relying solely on landfills, as has been done since time immemorial, the industry is developing an integrated waste management strategy. The objective is to minimize impact on the environment by employing all possible waste management technologies—especially reduction/reuse/recycling and incineration—in addition to landfills.

An integrated waste management strategy is required by law in many jurisdictions and is now being used in most North American communities. Most U.S. states, for example, have made recycling mandatory and have established goals for reducing waste per capita by 25% to 50% over a period of four to ten years. In Canada, a comprehensive waste reduction plan established in the province of Ontario in 1991 has the goal of reducing the amount of waste going to disposal by at least 50% per capita by the year 2000, compared to the base year of 1987. The goal is to be achieved through implementing the "three Rs": reduction (10%), reuse (15%), and recycling (25%). Some jurisdictions have set even higher goals; for example, Seattle is aiming to reduce waste going to landfill by 60% by the year 2000.

An integrated waste management plan follows the life cycle of consumer products from cradle to grave, seeking to maximize the useful life of the resources that are involved. A complete suite of elements that might be used in an integrated waste management system is illustrated in Figure 4.1, although any municipality may utilize only some of these.

1. *Source reduction:* The objective is to reduce the amount of waste that is created in the first place. This can be accomplished in a number of ways: purchasing products with minimal packaging; developing products that are more durable and easily repaired; substituting reusable products for disposable single-use products; or implementing tax and other economic measures to encourage producers to generate less waste and use fewer resources. For source reduction to have a significant impact, society needs to turn away from the current consumer preference for once-through, disposable, and limited-life products.

2. *Reuse:* This means reusing a product rather than discarding it. Examples include repairing old appliances, refurbishing old furniture, repairing old automobiles, and refilling bottles (as is currently done to a limited extent with soft drink, beer, and wine bottles).

3. *Recycling:* The objective is to convert waste materials into a usable form. Examples include the recycling of paper, metal and aluminum cans, glass bottles, and plastic, as is commonly being done through blue-box programs. Composting is another example of recycling: organic waste materials such as yard debris are converted into useful garden mulch.

4. *Incineration:* The process of burning rubbish has been employed for centuries because it destroys putrescibles (garbage that rots) and significantly decreases the volume of waste requiring disposal. When accompanied by the capture of heat energy for generation of electricity, district heating, or other purposes, it can be viewed as a form of recycling. In fact, it is often called the "fourth R." Incineration and its pros and cons are described in chapter 9.

5. *Treatment:* Wastes can be chemically or physically treated prior to disposal to improve their properties. Treatment can reduce the toxicity of waste, remove further useful components, and improve waste properties for disposal. For example, incinerator ash can be fixed in concrete to make it more leach-resistant prior to placing it in a landfill. Waste treatment is discussed in chapter 10.

6. *Landfilling:* Ultimately, some waste will remain, even after vigorous application of the other strategies. This waste will be placed in landfills, since no disposal alternatives are currently in common use. The technical details of landfills are described in

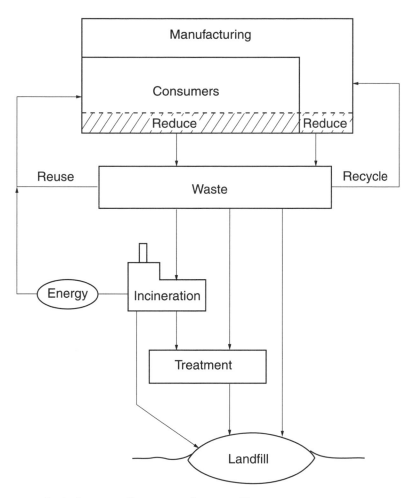

4.1 Block diagram illustrating the overall waste management system.

chapter 7, and a number of case histories are presented in chapter 11. But the landfill does not necessarily represent the end of the road. Although it is not commonly done, landfills can be exhumed (landfill mining) and their contents recycled and/or incinerated for energy.

The design and implementation of an integrated waste management system is a complex undertaking and must take into account many factors, including demographics, market accessibility for recycled products, and the availability of land and other resources.

A process should be established for communicating with the public so that their input can be obtained and incorporated into the master plan.

Comparative Cost Analysis—Apples and Oranges?

In designing an integrated waste management system, it is necessary to select a combination of various technologies. Currently, the choices are relatively restricted and are limited primarily to landfills, incinerators, and material recycling centres (which may include composting). Invariably, the financial cost of these facilities and their associated support programs plays a large, if not the largest, role in the selection process. Thus, no book on waste management would be complete without a discussion of finances.

Although some absolute cost values are included here, they will soon be out of date, so emphasis is placed on comparative or relative costs. In addition, "hidden" environmental costs are described.

Traditionally, the overall price of a specific facility is estimated by adding the costs of each major phase—siting, construction, operation over a 20- to 40-year lifetime, and decommissioning. It should be noted that, because of the NIMBY syndrome, siting has become a major fraction of the total cost. Allowance is also made for the sale of energy or recyclable goods, as appropriate.

When financial analyses are done, landfilling appears attractive and is often more economical than incineration or recycling. Typical landfill tipping fees (charged to haulers) range from $25 to $70 per tonne (metric ton), whereas the cost of waste incineration with energy recovery is in the range of $40 to $100 per tonne. Recycling costs vary significantly from place to place and time to time, depending on the vagaries of the recyclable commodity market, but they typically range from $40 to $90 per tonne.

The costs of a modern 1100-tonne-per-day energy-from-waste incinerator in Lancaster County, Pennsylvania, are presented in chapter 11. The capital cost, including siting and licensing, was $110 million. The operating cost is approximately $9.5 million per year, which is more than offset by electricity sales of about $12 million per year and the sale of ferrous metal (reclaimed from the ash) for $200,000 per year.

The costs for a modern recycling center in Guelph, Ontario, which includes composting, are presented in chapter 11. The cost of the center was $27 million, which included pilot studies, permits, the truck fleet, and land costs. Once full capacity is reached, the annual operating cost will be $3.7 million. This will be offset by revenue from sale of recyclables at about $4.4 million per year, yielding an operating profit of approximately $700,000 per year.

The low cost of disposing waste in landfills is due primarily to two factors. First, landfills involve relatively low technology; they are essentially large earth-moving exercises. By contrast, the other facilities—particularly incinerators—involve high technology, with control rooms, complex machinery, and high temperatures and pressures that must be carefully monitored. Second, recycle centers and incinerators require up-front capital, because they must be constructed before they can be used, whereas the capital cost of a landfill is spread over its lifetime. Landfills have lower ongoing operating costs owing to their relative simplicity. Furthermore, the financial outcome of incinerators and recycle centers is often complicated by the cyclic and unpredictable marketing of energy and recyclables. The bottom line is that, from a financial perspective, landfills frequently look like the way to go. Since we live in a capitalist, free market economy, does this not logically tell us to place all our waste in landfills?

Wrong! Traditional financial analyses omit several important factors.

First, landfills are effectively permanent facilities; that is, they will need to be monitored and maintained for centuries after they close. The full perpetual-care cost is seldom included in tipping fees. Even when post-closure costs are included, they seldom consider a time frame greater than 20 years. In contrast, incinerators and material recycle facilities are temporary structures: once their useful lifetimes (usually two to four decades) are over, they can be decommissioned and dismantled, and the land can be sold and put to other beneficial uses. Not only is a perpetual-care fund not required, but income is also gained through sale of the land.

But the most crucial factor is the cost of damage to the environment. It is becoming recognized that the value of Earth's natural ecosystems and the "services" they provide is not fully captured in commercial markets in the way that economic services and manufactured commodities are valued (Costanza et al., 1997). Ecosystems contribute to—indeed, are essential for—human welfare, and thus they represent part of the total economic value of the planet. Costanza et al. (1997) estimated the current economic value of the biosphere (most of which is outside the traditional market) to be approximately $33 trillion ($10^{12}$) per year, which is almost double the global gross national product! This value is seldom included in economic analyses, and as a result, environmental considerations are not adequately represented in policy decisions.

Material recycle centres are an excellent example of how standard economics do not include the value of our ecosystem. The arguments in their favor are compelling: recycle centres reduce the size of the entire waste management problem; they reduce demand for energy; they reduce pollution of land, air and water; and they conserve scarce raw materials. Yet recycling is not all that successful, and in many municipalities the cost of recycling programs is greater than that of landfilling.

To reflect properly the value of Earth's ecosystems, additional cost factors should be included in financial analyses of waste management facilities. For example, landfill costs should be increased to account for (1) permanent loss of land resource, (2) impairment of groundwater by leaking leachate, and (3) degradation of the Earth's atmosphere by gas emissions. Similarly, the cost of incinerators should include a factor for degradation of the atmosphere by gas emissions. In contrast, material recycle facilities may not be assessed any such "environmental" cost, since they do not appear to cause any appreciable long-term damage to ecosystems.

Unfortunately, ecosystem values are not readily included in financial analyses at present because commonly accepted methods and costs are not available. It is urged that economists undertake the necessary research to rectify this situation.

The inability to assign *quantitative* costs to environmental factors in waste management facilities does not mean that these factors should be ignored. Instead *qualitative* methods, such as weighting factors and preferences, should be applied.

How Much Recycling Is Achievable? Exploding the Myth

The argument is often put forth that there is no need for incinerators or landfills: recycling, source reduction, and other conservation efforts will eliminate all garbage. This approach seeks a return to the recycling and conservation ethic that was very successful in the early 1900s and during the two world wars.

It is questionable whether this idealistic goal can be achieved. But how much of our municipal waste actually can be recycled? Some idea of what is practically achievable is necessary for planning effective integrated waste management programs.

In consideration for our environment, we should be aiming for the ultimate target of no waste emplacement in landfills. As discussed in chapters 13 and 14, it is conceivable that this idealistic target may be met at some time in the future. At present, however, and in spite of much ballyhoo, very few communities in North America are exceeding 50% waste diversion, with most in the 15% to 40% range; the national average is approximately 21%. Can these numbers be increased? And by how much?

The Big Picture

Let us look at this question from a large-scale perspective. Figure 4.2 summarizes in graphic form the quantities of municipal solid waste generated in the United States, and how this has been managed from 1960 to 2000 (EPA, 1994).

The most significant point to be gleaned from Figure 4.2 is that the total amount of waste generated in the United States has increased each year and will continue to do so. The steady growth in garbage is due primarily to increasing population.

However, the rate of growth in total garbage slowed down after the mid-1980s, when the current era of the three Rs began, with the expansion of incineration, composting, and recycling programs. These programs are being vigorously promoted, so that by the year 2000 they will be diverting more than twice as much waste from landfills as they were diverting in 1985.

Motherhood, Apple Pie, and Recycling?

Although recycling is now accepted as the equivalent of motherhood and apple pie in terms of the "correct" approach to conserving resources, things are not always as simple as they seem.

A comprehensive life-cycle study of paper recycling in Europe yielded some surprising results (Virtanen and Nilsson, 1993). The study concluded that a balanced combination of paper recycling and energy recovery through waste incineration was most appropriate from an overall economic and environmental standpoint. A major factor is that incinerating paper, a renewable resource, saves fossil fuels, a nonrenewable resource. Furthermore, burning paper is cleaner than burning fossil fuels, which contain much higher concentrations of sulfur and other contaminants. Paper constitutes about 35% of household waste volume.

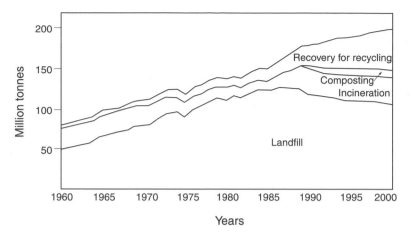

4.2 Municipal solid waste generated in the United States from 1960 to 2000.

The data from Figure 4.2 are summarized in Table 4.1 for the years 1985 and 2000. It is seen that the amount of waste going to landfill is reduced by only 23 million tonnes, or about 18% over this period, in spite of the vigorous growth in three Rs programs.

Significant gains are being made in waste diversion, but these serve largely to offset the growth in garbage generation caused by population increase. As a result, only a small net decrease in waste going to landfill has been achieved.

Dramatic future decreases in the amount of waste going to landfill seem unlikely, because North America's population is projected to continue growing, and because the easy methods for recycling have already been implemented, so further decreases will become ever more difficult. That is, the law of diminishing returns will start to play a role. Even if amazing new technologies for diverting waste from disposal were developed, it would still be many decades before our reliance on landfills could be broken.

The conclusion is inescapable: for the foreseeable future, the three Rs, important as they are, are not the complete solution to waste management. To make a really significant dent in total waste volume, additional methods are required.

The only technology that is currently capable of making such a contribution is incineration. Furthermore, even with vigorous recycling programs coupled with waste incineration, there will still be a need to dispose of wastes, albeit in smaller quantities, into landfills.

Table 4.1   Waste being recycled, incinerated, and landfilled in the United States of America (millions of tons).

Year	Recycled*	Incinerated	Landfilled	Total
1985	18 (11%)	18 (11%)	124 (78%)	160
2000	61 (31%)	36 (18%)	101 (51%)	198

*includes composting

Practical Experience

The recent era of the three Rs began in the mid-1980s. Initially, waste management plans were made with recycling targets of 50%; many communities even set targets of 60% to 70%. Since that time, experience has provided a factual basis for projecting maximum three Rs capacities. Currently, rates between 25% and 40% are typical in most communities. A few municipalities are achieving close to 50% waste recycling. Further gains, however, require increasingly more investment of resources as well as changes in consumer behavior. Almost no community has gone significantly beyond 50% waste recycling. It is clear that the law of diminishing returns has come into play. Even though 50% waste reduction through three Rs programs looks achievable, movement significantly beyond that figure appears neither technically achievable (for certain wastes), nor practical in terms of the costs that would be involved (Pratt, 1995).

Although a good start has been made since the mid-1980s in (re)introducing recycling as a major part of how we manage solid waste, there is still a long way to go. In fact, in some areas movement may have regressed. In 1964, 89% of all soft drinks and 50% of all beer in the United States were sold in refillable glass bottles. By 1993, refillable bottles made up only about 7% of the market, and were used in only ten states. This is in stark contrast to Europe: in Denmark, the use of nonrefillable beverage containers has been banned; in Finland, 95% of soft drink, beer, wine, and spirits containers are refillable; and in Germany, this number is 73%.

Another example of North American consumer resistance to waste reduction is the use of shopping bags. There has been considerable debate on whether plastic or paper bags are more environmentally friendly. Lost in the rhetoric is the simple fact that neither should be used—consumers should be bringing their own reusable bags. In the Netherlands, stores charge for supplying bags.

An Example of a Conserver Society

In Japan, the waste crisis was recognized as a national problem shortly after World War II. In response Japan developed the most comprehensive and successful recycling programs in the world. These can provide a realistic target of what is achievable.

Geography has imposed severe restrictions on Japan. It is an island nation with a small land mass and a large population. Approximately 1,400 people live on every square kilometer of habitable land, compared to only 50 persons per square kilometer in the United States. Thus, land is far too scarce a resource to waste on landfills.

With limited natural resources, Japan has tried vigorously to reduce its dependence on foreign raw materials by conserving and recycling to the maximum extent possible. For example, between 1980 and 1985, Japan imported 99.8% of its oil, 92% of its coal, and 65% of its lumber. Conservation is far ahead of what is practiced in North America.

In spite of its exemplary efforts, Japan has achieved recycling rates of only about 50%. Furthermore, Japanese officials believe that there is very little potential for further improvement; they feel that perhaps another 3% or so can be achieved (Hershkowitz & Salerni, 1987). (It should be noted that composting makes only a minimal contribution to Japanese recycling, with less than 0.2% of municipal waste being composted.)

To protect their valuable land resources from being used for landfills, the Japanese have turned to incinerators to complement their strong recycling programs. In the 25 years up to 1987, they built 1,900 incinerators, of which 69% (by capacity) generate energy (Hershkowitz & Salerni, 1987). In spite of concerns about dioxin emissions, air quality, and emissions from older plants, they have decided that incinerators must play a vital role in municipal waste management.

The overall goal in Japan is to reduce the amount of waste going to landfill. The success of their program is illustrated in Table 4.2, which compares the proportions of municipal waste recycled, incinerated, and landfilled in the United States, Canada, and Japan.

In North America, we need to emulate the Japanese approach. We must treat waste as a resource, a valuable economic commodity. We need to apply education, awareness, and "sanitary literacy" to achieve a strong waste reduction and recycling ethic.

Table 4.2   Comparison of U.S., Japanese, and Canadian municipal
waste management

	U.S. (1995) %	Japan (1987) %	Canada(1995) %
Recycled	21	50	21
Incinerated	20	34	4
Landfilled	59	16	75

The Bottom Line

In summary, waste reduction and recycling efforts are a critically
important part of waste management. They are not, however, the
complete solution: they will not eliminate all, or even most, of the
waste generated by society for many decades, if ever. Even Japan,
which has long faced enormous pressures to conserve, has achieved
only about 50% waste recycling.

Furthermore, the quantity of waste generated in North America
is continually increasing because of our growing population. Recycling alone simply cannot cope with this growing pile of garbage.

The only technology available today that contributes (together
with recycling) to a significant diversion of wastes from landfills is
incineration. Japan, and some communities in North America (see
the Lancaster County case history in chapter 11), have shown that
incineration and recycling are compatible and can be used together
in meeting the overall goal of reducing our reliance on landfills. The
two methods do not compete; they can be a compatible pair.

Discussion Topics and Assignments

1. How is solid waste managed in your community? What
   fractions are landfilled, recycled, composted, incinerated,
   and exported? Have any of these methods had negative
   impacts on local people? How could the overall system be
   improved?
2. What recycling is being done in your community? Is the
   program successful? How can the program be improved to
   (a) collect a greater fraction of recyclables, and (b) to make
   it more cost-effective?
3. A hypothetical community of 50,000 currently has no recycling programs. They wish to comply within three years

with the new state law requiring 30% diversion. How would you go about gathering data and preparing and implementing an integrated waste management plan?

4. Make a list of all the garbage that you discard over one week. How much of this did you recycle? How much more could have been recycled (with perfect recycling facilities)? Put the information on a spreadsheet, and compare it with data from classmates or colleagues. Repeat this exercise at other times of the year to see what effect the seasons have on your domestic garbage production.

5. A financial analysis of landfills should include the environmental cost of permanent loss of land, groundwater impairment, and atmospheric pollution. What factors should be considered in determining a dollar value for each of these? What dollar value would you assign for each?

Suggested Reading

Haun, J. W. 1991. *Guide to the Management of Hazardous Waste: A Handbook for the Businessman and the Concerned Citizen.* Golden, Col.: Fulcrum Publishing.

Hershkowitz, A., and E. Salerni. 1987. *Garbage Management in Japan: Leading the Way.* New York: Inform Inc.

5

RECYCLING AND COMPOSTING

Making a Molehill Out of a Mountain

Recycling, which includes composting, is the current rage. Almost every community in North America has established some kind of recycling program in the past few years. This chapter focuses on the science and technologies that are involved in recycling programs and explores what is needed to make these programs successful.

Recycling

This section describes the part of recycling that is associated with blue-box or streetside programs. It includes paper, cardboard, metal, aluminum, and plastics; composting is described in the next section.

Factors for Success

A successful waste recycling program relies on more than a systematic application of equipment and other resources. It also depends very significantly on attitude. It is vital that everyone participate.

To achieve a meaningful level of participation, some degree of legislative guidance may be necessary. In fact, studies have shown that mandatory recycling programs are much more effective than those run on a voluntary basis (Platt et al., 1991). Legislation or bylaws can also be used to

- stipulate that soft-drink, beer, wine, and other bottles be reused

- require the use of recycled material in manufacturing new products
- avoid excessive packaging
- reduce tipping fees for recyclable or compostable materials brought to designated drop-off sites
- set higher tipping fees for waste from which recyclables have not been removed
- ban the landfilling of certain substances, such as yard wastes

Public education is an indispensable part of an integrated waste management system. First, the public must be informed of the details that involve them: what days pickups are made, how to obtain recycle containers, what materials can be recycled, how they are to be sorted, and so on. This information can be disseminated by flyers, newsletters, ads in the local paper, features on local television channels, and telephone hot lines.

Second, an ethic of conservation should be instilled so that people will want to participate in three Rs programs. Methods of achieving this objective include videos and slide shows at schools, posters, buttons, and awards to businesses and groups that make outstanding contributions to recycling. These programs need to spell out clearly the benefits to the participants, to their children and grandchildren, and to the environment that arise from conserving resources in an environmentally responsible manner.

The operational aspects of waste management begin with collection of refuse from residential areas, apartment buildings, businesses, and industries. This is the interface between the waste generators and the waste management system. Today, most communities have some form of blue-box program to collect recyclables in addition to regular municipal waste. The collection program represents a significant cost as well as a formidable logistical exercise. It is essential that the waste generators—the customers—know what is to be put out and how it is to be sorted and separated.

Recycling Snippets

Recycling one aluminum can saves the energy equivalent of half a litre of gasoline, which is about 20 km of driving.

Recycling one glass bottle saves the energy equivalent of burning a 100-watt lightbulb for 4 hours.

→

Recycling Snippets (continued)

Recycling 54 kilograms of paper saves one tree.

30 million liters of oil are sold in Ontario each year. Less than 1 million liters are recycled. Where does the rest go?

Issues that must be addressed include the following:

- Does one truck pick up both regular garbage and recyclables, or are separate trucks required? Clearly, the costs escalate significantly if two different fleets are used, or even if the same trucks must be sent out on two different trips to the same area. Some communities have found a clever solution to this problem by attaching trailers to the backs of their regular garbage trucks. Specialized trucks are available which can be operated by one person who does both the driving and waste collection; these trucks have two or more separate internal compartments, each with its own compacting and tipping capability.
- Are garbage and recyclables picked up on different days? Clearly, the most economical solution is to combine the pickups. Aside from the expense of having the garbage/recycle trucks make two trips, it has been shown that public participation is lower if they must put out recyclables on a different day than regular garbage.
- Should garbage and recyclable pickup be on different schedules—say garbage weekly, and recyclables biweekly? Substantial savings can be realized if biweekly pickup is feasible. However, some studies (Platt et al., 1991) have shown that this leads to lower levels of recycling.
- How much separation of recyclables can be implemented by the waste generators? That is, will householders and businesses separate plastics, fine paper, cardboard, boxboard, metals, and so on without significant contamination?
- Can the truck pickup segregate the wastes in the same categories as is done by the generators? Separation of recyclables generally requires side-loading trucks. If the trucks cannot provide separation, then the materials recovery facility will need to include equipment for separating cans, bottles, plastics, etc.
- A key to a successful recycling program is to ensure that apartment complexes and businesses also participate. This

requires some innovative coordination, because apartments generally have space limitations, and businesses are often serviced by private-sector waste companies rather than by municipal waste pickup.

A study of 17 communities that have established successful recycling programs showed that 60% of them are diverting 40% or more of their waste from landfills. The study found that the following factors were important in achieving such high diversion rates (Platt et al., 1991):

- mandatory participation in recycling
- comprehensive composting programs
- recovery of materials not only from residences but also from apartments and commercial and institutional establishments
- targeting a wide range of materials for recovery
- providing economic incentives for materials recovery, such as increased tipping fees for nonseparated refuse
- weekly pick-up of all materials
- provision of adequate containers for recyclable materials
- implementing education and publicity programs

Materials Recovery Facilities (MRFs)

Materials recovery facilities (MRFs, pronounced "murfs") are the most critical component of a modern waste management system. The efficiency and flexibility of the MRF will control the percentage of waste that a municipality can recycle. Recyclable wastes that have been picked up at curbside are brought to the MRF, where the various metal, glass, paper, and plastic components are separated, packaged (usually into bales), and stored until they can be shipped. MRFs generally include a drop-off area so that individuals or companies can bring in presorted recyclable materials. Truck access to the building and traffic control are important so that materials can be dropped off and the final baled products can be picked up.

If considerable sorting is done in the trucks that do the curbside pickup, the materials recovery facility may need to provide only minimal additional sorting capability, although some may still be necessary to remove contamination and ensure that a sufficiently "clean" product is supplied to the purchaser. If little sorting is done in the curbside pickup, then more sophisticated sorting equipment will be necessary.

The main operations and equipment at a MRF include the following (Tchobanoglous et al., 1993).

Conveyor Belts   Various types and sizes of conveyer belts are ubiquitous at MRFs and are the workhorses for managing and transporting materials. They are also used as "moving tables" to allow manual sorting.

Size Reduction   The objective of size reduction is to produce a final product that is reasonably uniform and considerably reduced in size from the original, so that it can be more easily shipped and processed. Shredding, milling, and grinding are terms used to describe size reduction. Commonly used size reduction equipment includes the hammermill, which shatters brittle materials like glass using the impact of a number of hammers; the shear shredder, which uses two opposing counteracting blades to cut ductile materials; and the tub grinder, which is essentially a mobile hammermill shredder that is used to shred yard wastes and construction debris.

The most important operation at a MRF is the separation of the various recyclables from the overall mass of material delivered to the facility. Many different methods are used; the most commonplace are described below.

Size separation.   Different types of screens are used to separate mixtures of materials of different size.

Vibrating Screens are generally horizontal or slightly inclined, forming flat surfaces on which material is placed. The screens are vibrated so that materials of a specific size fall through the openings in the screen. Most screen systems allow several screens to be used, one above the other.

Trommel Screens are very versatile and consist of a large-diameter screen in the shape of a cylinder that rotates on a horizontal or inclined axis.

Disc Screens consist of sets of parallel, vertical, interlocking, rotating discs. Undersize materials fall between the discs, while oversize materials are carried along the top.

Density separation.   Density separation is used to separate materials on the basis not only of density but also of their aerodynamic properties. The principle is similar to that of winnowing grain.

Air classifiers are used to separate lighter materials such as paper and plastic from denser materials such as metals and glass. Mixed material is introduced into a fast-flowing air stream so that the lighter fraction is carried away into a cyclone separator or onto a different conveyor belt where it is collected, and the heavier fraction drops downward to its collection area.

Stoners, originally used to separate stones from wheat, are used in MRFs to separate heavy grit from organic material in trommel underflow (i.e., undersize) streams. A stoner consists of a flat deck with small air holes that is inclined slightly to the horizontal. The deck vibrates in a straight line in the upslope direction while an air stream is blown up through the holes. When material is introduced onto the platform, the light material floats on the air stream and flows down slope. The heavier material settles on the platform and is vibrated upslope to its collection area.

Magnetic and electric field separation.   Magnetic separation is used to extract ferrous metals, such as cans, from the waste stream. An electromagnet is generally used in conjunction with conveyor belts. Figure 5.1 illustrates how magnetic separation operates.

Eddy current separation is used to separate nonferrous metals, primarily aluminum, from plastics and glass. In this method an alternating electromagnetic field is used to induce currents in the aluminum. The induced currents create an electromagnetic field that is opposed to the inducing field, thus repelling the pieces of aluminum away from the inducing field (in contrast to magnetic separation, where ferrous metal is attracted to a magnetic field).

Manual sorting.   MRFs also rely on a considerable amount of manual sorting, even when they employ state-of-the-art separation technology. The current trend is to integrate manual and mechanical sorting of wastes because manual sorting provides final materials of the highest quality—that is, with the least amount of contamination.

After Sorting

In almost all cases, the end products produced by the various product streams in a MRF are compressed in compactors to reduce volume and packaged in bales which can be handled by forklift trucks.

Weigh scales are an integral part of a MRF. They are used to weigh trucks before and after bringing in loads and to weigh the various streams produced by the sorting processes to ensure proper accounting of the various marketable materials and the waste that must go to landfill.

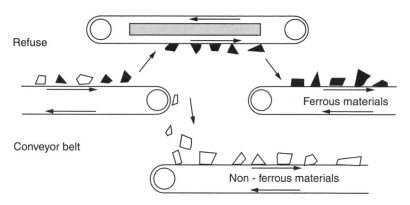

5.1   Magnetic separation of ferrous and nonferrous materials.

The materials recovery facility includes a storage area for the baled materials. A large storage space provides greater flexibility; it also allows for long-term storage should prices for a commodity be temporarily in the doldrums.

A flow chart illustrating the movement and separation of materials in a hypothetical materials recovery facility is shown in Figure 5.2, and a case history of a state-of-the-art recycling center is presented in chapter 11.

Marketing

Marketing is a fundamental requirement of modern recycling: the entire system would fail if the recovered materials could not be sold, or if they did not command a reasonable price. This is a complex business. Specialized markets must be found, and commodity prices can vary dramatically over time. It is not easy for municipalities, which generally have no experience in this kind of marketing, to master all of the intricacies of the trade. In many cases, it may be appropriate to contract the marketing of recovered materials to a firm that specializes in this challenging business.

Composting

Composting is a specialized part of recycling in which organic wastes are biologically decomposed under controlled conditions to

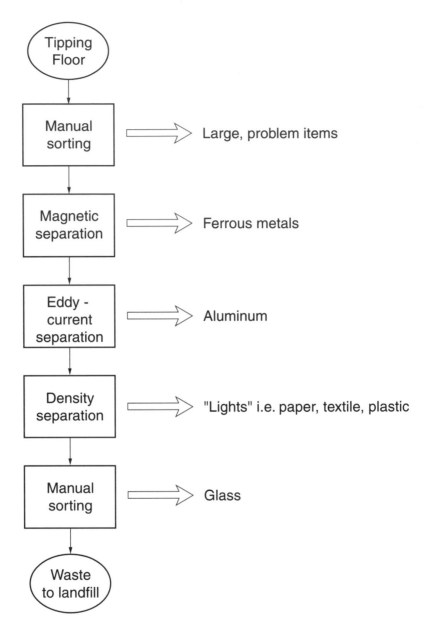

5.2   Simplified materials flow for separating recyclables in a materials recycle facility.

convert them into a product that can be applied to the land benefi-
cially and without adverse environmental impact. The composting
process should destroy pathogens, weed seeds, insect eggs, and
other unwanted organisms. Adding compost can lighten heavy soils,
improve the texture of light soils, and increase water retention capacity. Composting is a natural process that has been used in an
organized fashion to deal with garbage since at least the early 1900s
(*Journal of Waste Recycling*, 1991).

Composting is an important component of a modern integrated
waste system for one very simple reason: in North America we generate a considerable amount of yard waste and other organic wastes
that are readily compostable. Studies have shown that a significant
fraction of municipal solid waste consists of yard wastes, ranging from
5% to 20% by weight, with a typical value of 18%. Thus, composting
can make a significant contribution to waste diversion. Furthermore,
composting is a relatively low-technology and low-cost process that
can be readily established by most communities.

Generally, only materials of biological origin—such as leaves,
paper, wood, and non-meat food scraps—are suitable for composting.
Synthetic organic materials, particularly plastics and rubber, are seldom compostable.

The Composting Process

In the composting process, microorganisms break down complex
organic molecules (proteins, amino acids, lipids, carbohydrates, and
cellulose) into simpler ones (mostly cellulose and lignin). The
microorganisms require an aqueous or moist environment and oxygen. The exothermic reaction is depicted below:

$$\text{Complex molecules} + O_2 + \text{microorganisms}$$
$$\rightarrow \text{compost} + \text{new cells} + \text{dead cells} + \text{heat} + CO_2 + H_2O + NO_3 + SO_4$$

The activity of living organisms, which make up about 5% to 10%
of the organic material, releases the energy and nutrients stored in
the tissues of the plant and animal residues in the starting compost
material. There are several different kinds of organisms, and each
has a specific substrate on which it works. An entire food chain
develops during the compost process:

- Microorganisms such as bacteria, actinomycetes (slime
  molds), fungi, and algae break down the bulk of organic

material. Their population, commonly referred to as the "microbial biomass," is most crucial to the process.

- Protozoa, nematodes, and some other small organisms such as mold mites (*Acari*) and springtails (*Collembola*) feed on the microorganisms.
- Beetles and other insects feed on the mold mites, springtails, and other small organisms.
- Larger organisms such as earthworms, flatworms, centipedes, millipedes, snails, slugs, and sowbugs feed on the decaying plant materials. They speed up the compost process by mixing the materials and reducing the size of particles.

The carbon/nitrogen (C/N) ratio is the most important measure of nutrient balance in the compost. Microorganisms use carbon as a source of energy, and both carbon and nitrogen are used for building cell structures. The C/N ratio declines as the composting process proceeds. More carbon is required than nitrogen; a typical final C/N value is approximately 22:1 (MOE, 1991).

The C/N value determines how the finished compost affects the soils to which it is applied. If C/N is greater than 25:1, the microorganisms in the compost will compete with the crops for available nitrogen. At compost levels below 20:1, the energy source, carbon, is less than needed for conversion of nitrogen into proteins. In this case, the compost microorganisms remove excess nitrogen as ammonia, denying it to plants and thus inhibiting plant growth. The C/N ratio in compost can be controlled by adding either highly nitrogenous materials like grass clippings and green vegetation, or highly carbonaceous materials like hay and dry leaves.

Home Composting

Composting programs take two distinct forms. The first employs home composters, usually plastic bins or barrels with a capacity of about 200 liters (see Figure 5.3). These are supplied to homeowners, often on a subsidized basis, and should be accompanied by instructions on what and how to compost. Additional support can be provided by telephone hot lines and by volunteer programs in which experienced composters provide assistance and advice to beginners. Although home composting programs are feasible only in suburban areas, they are very effective because waste is diverted at source and no pickup or treatment by the municipal system is required.

5.3   Home composters.

Central Composting Facilities

Even with a successful home composting program, a central composting facility can make a valuable contribution to waste reduction. The central facility can service apartment buildings, businesses, and neighborhoods where home composters are not feasible; in addition, it can treat leaves in the fall and Christmas trees in winter. Incentives should be developed to ensure that landscaping firms, significant generators of yard wastes, drop off their wastes at the central facility.

An important part of planning a central composting facility is obtaining regulatory permits, including communicating with local groups that may be affected by the facility. A relatively large parcel of land is required, and this is often located at the municipal landfill: land is available, garbage/recycle trucks come there anyway, infrastructure such as weigh scales and wood shredders is available, and the final compost can be used for landfill cover if no other markets are available.

The processing of organic materials prior to composting includes shredding to break bags, reduce size of materials such as Christmas trees and large wood pieces, and ensure a relatively uniform material; and sorting to remove contaminants such as plastic bags.

Composting facilities, though a relatively low technology, still require careful planning and resources. Generally, three basic systems are used: the windrow, static pile, and in-vessel methods (Tchobanoglous et al., 1993). These are described below. The windrow and static pile methods are the most popular license they require minimal capital investment and the decomposition process occurs aerobically (in the presence of oxygen). In aerobic composting (versus anaerobic composting, in the absence of oxygen) far less odor is generated, and temperatures reach higher levels, generally in the 40° to 60°C range, which not only kill most pathogens but also destroy weed seeds.

Windrow composting.   This is one of the oldest and simplest methods of composting. A typical windrow system consists of long rows of organic material, about 1.8 to 2.1 meters high and 4 to 5 meters wide at the base. Actual dimensions vary and depend largely on the equipment available to place and manipulate the piles.

To ensure aerobic conditions and maintain temperatures, the windrows are turned at regular intervals, usually once or twice a week. A moisture content of 50% to 60% must be maintained. Although bulldozers and front-end loaders can be used, specialized turning machines have been developed that are more efficient and can add water at the same time (Fig. 5.4). Proper aeration is important because it prevents anaerobic conditions, which lead to odor. A temperature of at least 55° C should be maintained for a minimum of two weeks to ensure destruction of pathogens. The composting period lasts about four or five weeks; the compost is usually cured for an additional two to eight weeks to ensure that it is completely stabilized.

Aerated static pile composting.   This method can be used to compost a wide variety of organic materials, including yard wastes and separated municipal waste. The materials are laid out in long piles similar to windrows. A layer of screened compost is often placed on top of the pile to control odor and provide insulation. A network of perforated piping is either placed at the bottom of the pile or embedded in the flooring below the pile. Air is introduced by blowers into each pile through the pipe network so that aerobic decomposition occurs. Airflow rates are controlled to maintain the temperature at the desired level. In modern facilities, all or most of the system is enclosed to allow better processing and odor control. Al-

5.4   A self-propelled windrow-turning machine (courtesy of
Midwest Bio-Systems, Tampico, Illinois).

though the method needs more complex equipment than windrow
composting, it does not require turning the material, it minimizes
odors, and it provides better control of the process. An aerated static
pile composter is illustrated in Figure 5.5.

In-vessel composting.   In this method, the materials to be composted
are enclosed in a container or vessel. Vessels of various shapes are
used, but they are generally of two basic types: plug flow or dynamic. In the former, the materials move through the vessel without agitation; in the latter, the materials are agitated or mixed during the composting. Air and water are added to the vessels in a
well-controlled manner. Typical in-vessel composting systems are
shown schematically in Figure 5.6, and an actual system is shown
in Figure 5.7. Detention (processing) times in in-vessel composters
are about 1 to 2 weeks, followed by a 4- to 12-week curing period.
In-vessel composters are gaining popularity because they offer good
process and odor control, shorter composting time, and lower labor
costs, and they can deal with food wastes. In particular, they can
be set up in cities to service facilities such as hospitals or large office complexes.

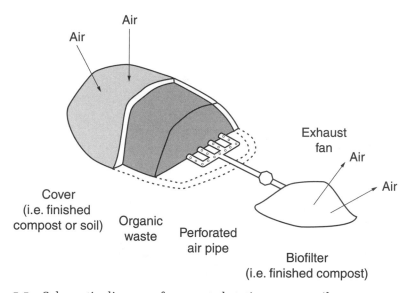

5.5 Schematic diagram of an aerated static compost pile.

(a)

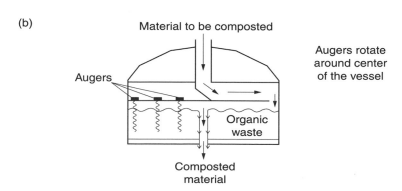

5.6 Schematic views of typical in-vessel composting systems: (a) horizontal plug flow composter; (b) mixed vertical flow composter.

5.7   A flow-through in-vessel composting system with internal
mixing and computer control of air volume and temperature,
situated at a municipal hospital (courtesy of Wright Environmental
Management Inc.).

Treating and Marketing Compost

Regardless of the composting method, the final product is usually
passed through a screen to remove oversize pieces and foreign materials as well as to ensure that it has a good marketable appearance.

Because of the high moisture levels that must be maintained
during composting, considerable leachate may be produced. Depending on the types of materials being composted and the permitting requirements, a leachate collection system, a water treatment
facility, and an environmental monitoring program may be necessary. Finally, the finished compost must be sold to a market, so a
storage, packaging and loading area forms a key part of the facility.

Discussion Topics and Assignments

1. Arrange for a tour of your local materials recycling facility. Make a material flow chart of the facility. Ask them
   what their main problems are and brainstorm ways of overcoming them.
2. Develop a flow chart for processing (separating) a waste
   stream that consists of paper (43%), cardboard (12%), plas-

tics (13%), glass (14%), aluminum (6%), and ferrous metal (9%). Assuming 100 tonnes/day, 5 days/week input, sketch a layout of the plant.

3. How much land is needed to establish a composting center if a community collects 100 tonnes/day, 5 days/week, of compostable material. Assume that windrow composting with mechanical turning will be used. Each windrow is 5 meters wide and 150 meters long, with an angle of repose of 1 to 1. The material has a density of 330 kg/m³ before water is added. The active composting period lasts 30 days and the curing period is four months. Don't forget a staging area. Repeat the calculation for an active period of 21 days and a curing period of three months.

4. Prior to composting, leaves, with a C/N ratio of 55:1, are to be blended with activated sludge from a sewage treatment plant, with a C/N ratio of 6.6:1, to yield a combined C/N ratio of 25:1. In what proportions should they be combined if the moisture contents of sludge and leaves are 75% and 50%, respectively, and the nitrogen content of sludge and leaves are 5.6% and 0.8%, respectively?

5. What is the C/N ratio of $C_{60}H_{100}O_{40}N_2$?

6. Investigate what is being done in your community to recycle the following two special waste types: (a) construction and demolition waste; and (b) food scraps from restaurants and cafeterias (as opposed to yard waste). Is there room for improvement? How?

Suggested Reading

Journal of Waste Recycling. 1991. *The Biocycle Guide to the Art and Science of Composting.* Emmaus, Pa.: JG Press.

Platt, B., C. Doherty, A. C. Broughton, and D. Morris. 1991. *Beyond 40%: Record-Setting Recycling and Composting Programs.* Washington, D.C.: Institute for Local Self-Reliance, Island Press.

Tchobanoglous, G., H. Thiesen, and S. Vigil. 1993. *Integrated Solid Waste Management: Engineering Principles and Management Issues.* New York: McGraw-Hill.

WASTES
Know Your Enemy

When Sherlock Holmes solves a mystery, he studies the strengths, weaknesses, foibles, egos, sensitivities, and other traits of the villains. It is the same with wastes: a detailed understanding of their characteristics is fundamental to being able to manage them properly. To determine the size of a disposal facility, we must know the volumes and rate of generation of waste. A MRF cannot be designed unless it is known what recyclables are contained in the waste stream. A knowledge of the physical and chemical nature of waste allows engineers to select landfill construction materials that will be compatible with the waste. We must understand the toxic and hazardous components in order to design the facility to endure for a period of time commensurate with the hazardous lifetime of the waste.

Because of the incredibly large number of existing waste compounds, it is useful to categorize them. Unfortunately, there are no well-established categorization systems in place. We will describe wastes using two main classification systems, and then we will describe their most important characteristics. The first system is a functional one; that is, the wastes are classified by generator. The second is a classification by chemical type.

Classifying Wastes by Generator

This somewhat arbitrary system combines different kinds of waste primarily by the group or industry that generates the waste. These waste types include:

- municipal wastes
- industrial wastes
- hazardous wastes
- radioactive wastes

This is a convenient classification because each of these waste classes is generally managed and disposed of as a group. In addition, substantial volumes of waste are generated by the mining and agricultural sectors; these are not discussed in this book.

Municipal Solid Waste

Municipal solid wastes, as the name implies, are produced by the everyday activities in a community. They arise from the following sources:

- residential—houses and apartments
- commercial—stores, restaurants, office buildings, service stations, etc.
- institutional—schools, courthouses, hospitals, etc.
- construction and demolition—construction sites, road repair, building demolition, etc.
- municipal services—street-cleaning, garden and park landscaping, wastewater treatment, etc.

We are a wasteful society. Every person in North America generates approximately 2 kilograms of garbage each day. Given that there are approximately 300 million people in the United States and Canada, this yields over 200 million tonnes per year—a gigantic amount of garbage.

Just what does this waste consist of? Until recently, not very much was known about the garbage we produce. It is hardly surprising that the malodorous field of garbology has not attained the popularity of rocket science, oil exploration, or brain surgery. The situation improved dramatically in the early 1970s when the Garbage Project was launched at the University of Arizona (Rathje & Psihoyos, 1991; Rathje & Murphy, 1992). Motivated by the study of archeology, combined with modern demographics and lifestyle, the project undertook an ambitious study of today's landfills. In excavating into and performing detailed cataloging of the contents of landfills, the Garbage Project has brought a certain degree of glamor and excitement to garbology. Their studies, which have been featured in *National Geographic* and on television specials, have unearthed fascinating information about our lifestyles and our modern throw-away society.

The Garbage Project exhumed 11 U.S. landfills and showed that municipal garbage consists of the following components, by volume (see Fig. 6.1):

50% *Paper*, including packaging, newspapers, telephone books, glossy magazines, and mail-order catalogues. Paper is almost 100% cellulose, a carbohydrate which is highly combustible.

19% *Miscellaneous*, including construction and demolition debris, tires, textiles, rubber, and disposable diapers.

13% *Organic materials* such as wood, yard waste, and food scraps.

10% *Plastic*, including milk jugs, soda bottles, food packaging, garbage bags, and polystyrene foam. Plastic materials are classified into the following seven categories (see Fig. 6.2):

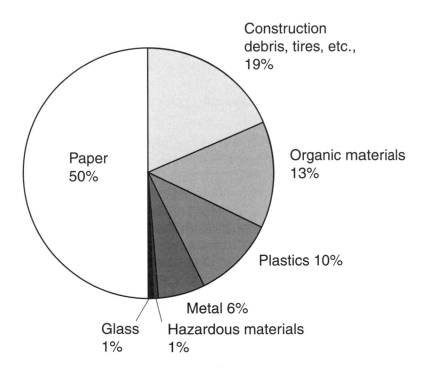

6.1   Composition of municipal solid waste.

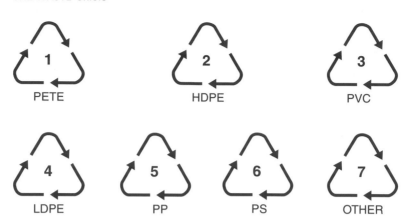

6.2   Symbols used for different types of plastic.

- Polyethylene terephthalate (PETE/1)
- High-density polyethylene (HDPE/2)
- Polyvinyl chloride (PVC/3)
- Low-density polyethylene (LDPE/4)
- Polypropylene (PP/5)
- Polystyrene (PS/6)
- Other multilayered plastics (7)

6%   *Metal*, including iron as well as aluminum and steel food and beverage cans.

1%   *Glass*, consisting of beverage bottles, food containers, and cosmetic jars.

1%   *Hazardous materials* such as pesticides and oven cleaners.

We can see that a large fraction of a landfill's contents is potentially recyclable. For example, approximately one-half of a typical landfill is comprised of paper. The single most abundant item in landfills is newspapers, accounting for about 18% of the space. Telephone directories are the fastest-growing paper component of municipal waste. An interesting observation of the Garbage Project was that disposable diapers, which are used by 85% of American babies and which are perceived by many to be a serious concern, actually account for only about 0.8% of landfill volume.

An obvious characteristic of municipal waste is its heterogeneity. Municipal landfills are anything but uniform, with an enormous range of items. Metals, for example, consist of items as disparate as refrigerators, car axles, paint cans, metal stakes, and pop cans.

Throw-Away Society

Each year U.S. consumers throw away:

- enough tires to encircle the Earth almost three times
- about 18 billion disposable diapers; if laid end to end these could stretch to the Moon and back seven times
- 10 million computers
- 1.6 billion ball-point pens

About 16% of municipal landfills are also used for co-disposal of sludge from sewage treatment plants. This is not surprising, since sewage treatment plants and landfills are generally owned by the same jurisdiction, permitting good cooperation between them. Landfills are generally a choice of last resort for sludge disposal, used only if no worthwhile use for the sludge can be found, such as fertilizer for farmers' fields or fuel for incinerators.

As described later in this chapter, municipal wastes are not just composed of benign items such as grass clippings and old newspapers. They also contain used motor oil, batteries, solvents such as oven cleaners and paint thinners, pesticides, tires, and much more.

Industrial Waste

We produce garbage not just in the home; the industries that supply our consumer goods also generate large amounts of solid waste. These wastes are byproducts of the manufacturing or chemical processes used by different industries, and the amount produced is approximately four times greater, by weight, than the amount of municipal solid waste. Even larger quantities of solid waste are produced by the mining, oil and gas, and agriculture sectors (which are not discussed in this text).

Industrial wastes are generally quite specific, depending on the manufacturing processes involved. For example, an integral part of the steel-making industry is the conversion of coal to coke in high-temperature coking ovens. This process emits a large amount of gas and dust. The dust is captured in baghouses or electrostatic precipitators (see chapter 9) and is generally disposed in dedicated landfills at or near the steel mill.

To give some idea of their esoteric nature, here are a few examples of industrial waste streams:

- Brine purification muds from the mercury cell process in chlorine production
- Distillation or fractionation column bottoms from the production of chlorobenzenes
- Emission control dust and sludge from the production of electrical power from burning coal
- Residue from the use of activated carbon from decolorization in the production of veterinary pharmaceuticals from arsenic or organo-arsenic compounds.

If any waste stream is hazardous, as defined by the presiding regulatory agency, then that waste must be disposed of into a licensed hazardous waste landfill. Nonhazardous wastes are either incinerated or placed into landfills, which are generally dedicated for that purpose and located at or near the factory producing the waste. These landfills are, for practical purposes, similar to municipal landfills (landfill design is described in chapter 7).

Hazardous Waste

Although hazardous waste is produced by a variety of different generators, it is grouped into one category because regulatory agencies require that it be separated from municipal and industrial waste and placed into separate, more rigorously designed landfills. For example, municipalities hold household hazardous waste days to separate this component from other municipal waste. Regulations require industries to analyze their waste, and the hazardous component must be removed by licensed carriers to a licensed hazardous landfill.

Although hazardous waste is defined in different ways by different jurisdictions, it is generally considered to be waste that is toxic or hazardous to humans or to the environment. The U.S. Resource Conservation and Recovery Act (RCRA) provides a good example of a definition created by committee. RCRA defines hazardous waste as

> a solid waste or a combination of solid wastes that, because of its quantity, or physical, chemical, or infectious characteristics, may cause, or significantly contribute to, an increase in mortality or an increase in serious irreversible, or incapacitating reversible, illness; or pose a substantial present or potential hazard to human health or the environment when

improperly treated, stored, transported, or disposed of, or otherwise managed.

Solid wastes may be deemed hazardous under RCRA if they exhibit one or more of the following four characteristics:

- Ignitability
- Corrosivity
- Reactivity
- Toxicity

About 15% of industrial waste (Oweis & Khera, 1998) and about 1% of municipal refuse is hazardous. It is estimated that about 40 million tonnes of hazardous waste are generated in the United States each year (Henry & Runnals, 1989). In Canada, it has been estimated that approximately 2.3 million tonnes are produced annually (Apogee Research, 1995).

Typical household hazardous wastes include everyday products such as nail-polish remover, batteries, oven cleaner, used motor oil, degreasers, fuel additives, paints, stains, varnishes, turpentines, pesticides, herbicides, and fertilizers. For example, a seemingly innocent bottle of fingernail polish contains the toxins xylene, toluene, and dibutyl phthalate.

The Garbage Project turned up medical products, including used needles, in their excavations. They found that almost 1% by weight of all garbage coming from households can be regarded as hazardous, based on the Environmental Protection Agency's definition (Rathje & Murphy, 1992). Studies in Seattle and Montreal found that household hazardous wastes comprised 0.4% and 0.7%, respectively, of their municipal solid waste (Apogee Research, 1995). Although these are not large percentages, they actually represent a significant quantity of hazardous material. For example, in 1995 Americans produced 180 million tonnes of municipal trash; 1% of this, or 1.8 million tonnes, was hazardous waste, most of which wound up in landfills. Although household hazardous waste programs are becoming more popular, it is estimated that they intercept less than 4% of hazardous waste (Apogee Research, 1995).

Garbage Project studies have shown that hazardous wastes are not a new phenomenon but have been produced at about the current rate for more than 50 years (Rathje & Murphy, 1992). Thus, both old and new landfills contain significant amounts of hazardous materials, which will eventually leak into the groundwater.

After being segregated, hazardous wastes must be disposed into specially licensed hazardous waste facilities or into special hazardous waste cells located at regular landfills. Generally, special landfills designed for hazardous wastes incorporate particularly good geologic features, such as very thick and impermeable clay layers, and/or extra engineered barriers, such as synthetic plastic liners.

Hazardous wastes can be classified in a number of ways. For example, the University of California at Davis Waste Classification System, which is used widely, recognizes 116 individual waste types. They are grouped into the following 14 categories based on similar physical and chemical properties:

- Organic sludges and still bottoms (no oil)
- Solvents and organic solutions (halogenated and non-halogenated mixed)
- Oils and greases (waste lube oil, etc.)
- Oil/water mixtures
- Organic and oily residues (gasoline and organic food waste oil)
- Heavy metal solutions and residuals
- Miscellaneous chemicals and products (organic and inorganic chemicals, salts, infectious)
- Paint and organic residuals
- Aqueous solutions with organics (90% water)
- Anion complexes (cyanide, sulfide, other complexes)
- Sludges and inorganic residuals (ash, spent catalyst)
- Pesticide and herbicide wastes
- PCB wastes
- Clean-up residuals (largely contaminated soil or wastes absorbed onto a solid matrix)

Since many of these waste forms are liquid, they may wind up being discharged into sewage treatment plants, or they may be treated chemically rather than being sent to landfill. Even if specialized treatment processes are used, there are usually some solid or sludge byproducts left which are sent for disposal in landfills.

Radioactive Waste

Radioactive wastes are a specific type of industrial waste. The main generators are electricity-producing nuclear reactors. Some radioactive wastes are also produced by research, medical procedures, and specialized industrial processes.

Radioactive wastes contain atoms with unstable nuclei which undergo radioactive decay whereby they give up energy and transform to a more stable form. The energy emitted can be in the form of particles and electromagnetic radiation—usually either gamma rays (a form of energy), beta particles (electrons), or alpha particles (the nuclei of helium atoms, consisting of 2 neutrons and 2 protons). These are called "ionizing radiation" because they produce electrically charged particles, known as ions, as they pass through matter. The main adverse effect of radiation on humans is cancer, which may not become evident until many years after exposure to radiation.

Radioactive compounds differ in one fundamental way from nonradioactive compounds: they can have influence at a distance. This is due primarily to gamma radiation, which can penetrate matter. Thus, radioactive compounds can cause adverse health effects at a distance as well as by direct contact and ingestion. This characteristic also has a positive side, however, since it allows radioactive materials to be detected using Geiger counters and other monitoring devices, thus avoiding sampling and chemical analyses, a more costly and time-consuming operation.

High-level radioactive wastes are generated in nuclear power plants by splitting, or fission, of uranium atoms in a controlled nuclear reaction. The fission process releases an enormous amount of energy in the form of heat, which is used to boil water into steam that drives a turbine generator to create electricity. The used uranium fuel that comes out of a reactor is highly radioactive and must be managed with great care. A few hundred different radioactive radionuclides are formed and emit such intense radiation that shielding is required to handle the fuel bundles. The radiation decays—decreases over time—so that after approximately 300 years the fuel bundles have about the same radioactivity as the original uranium ore from which they were made. At this time, they can be handled without shielding.

Major programs are under way in several countries to develop safe disposal techniques for high-level nuclear wastes. In virtually all cases, the proposed disposal method is by burial deep in stable, carefully selected geologic formations. In the United States a site is being explored in Yucca Mountain, Nevada, which has extensive deposits of tuff, a volcanic rock (Reynolds, 1996). This site is in an arid region, and the groundwater is very deep—about 600 meters below the surface. If the site proves acceptable, a repository will be constructed at a depth of 300 meters, approximately midway between the surface and the groundwater table.

In Canada (Dormuth & Nuttall, 1987) and Sweden, the intention is to dispose of spent fuel by emplacement into the dense, hard, crystalline rocks of the Precambrian Shield at a depth of about 500 meters. These rock formations are among the oldest (more than 600 million years old) and most stable in the world. In addition, a number of engineered barriers will be used. The spent fuel, encased in titanium or copper containers, is to be lowered into cylindrical holes in the floor of underground tunnels. The remainder of the emplacement holes will be filled with compacted bentonite clay. The tunnels and shafts will be sealed with a mixture of bentonite clay and sand. The bentonite clay buffers, together with the geological formation, form an integrated multi-barrier system to contain the wastes.

Low-level radioactive wastes consist largely of ordinary trash and other items that have come into contact with radioactive materials and have become slightly radioactive themselves. They include plastic gloves and other protective clothing, machine parts and tools, test tubes, syringes, filters, and general residues and scraps that have been contaminated. Low-level radioactive waste is created primarily by the operation of nuclear power plants, although some is also generated by the medical profession and by certain industrial applications such as sterilization of medical and other equipment. Hundreds of different radionuclides can occur in low-level waste. Although the level of radioactivity of these wastes is considerably lower than that of high-level wastes, they must still be managed with care to prevent harm to the environment. Low-level radioactive waste decays and becomes less hazardous as time passes. Because of their low radiation intensity, little or no shielding is needed to handle low-level radioactive wastes.

A number of different methods are currently being used for disposal of low-level nuclear wastes. In Sweden, the wastes are being placed in mined caverns below the sea floor (see chapter 11). In France, they are being entombed in heavily engineered surface vaults which feature thick concrete barriers. In the United States, trenches located in thick clay formations are being used at a site near Barnwell, South Carolina.

Classifying Waste by Chemical Composition

Another way of classifying wastes is by their chemical composition. This is a more rigorous categorization than that by disposal type because it provides a quantifiable description as well as information about the properties of the wastes. The main categories are:

- Organic compounds
- Inorganic compounds
- Microbiological materials

Organic Compounds

Organic chemistry is the study of carbon compounds, particularly those in which carbon atoms combine with hydrogen, oxygen, and nitrogen. Such compounds are formed naturally by living organisms; they can also be made synthetically.

Important properties of organic compounds are specific gravity, the degree to which an organic compound sinks or floats in water; solubility, or how readily it can be dissolved in water; volatility, the degree to which a chemical evaporates in air; adsorption, the tendency to stick to the surface of other matter, such as soils; and degradation, the tendency to decompose into simpler molecules. A primary characteristic of organic compounds is that they are capable of being destroyed by combustion. The following are the main families of organic compounds that are of interest in waste management.

Chlorinated (halogenated) solvents.   These are generally soluble, volatile, poorly adsorbed, and slowly degraded. Typical chlorinated solvents include trichloroethylene, methylene chloride, and carbon tetrachloride. They are generally used as cleaning solvents (paint removers, metal degreasers), blowing agents in foams, and in the manufacture of chlorofluorocarbons (CFCs), along with many other industrial uses. Because of the attached chlorine or other halogen atom, these compounds are chemically reactive and damaging to human health and the environment. For example, carbon tetrachloride is highly toxic and can damage the gastrointestinal tract and the central nervous system; methylene chloride is a carcinogen and a narcotic.

Non-chlorinated solvents.   These are generally soluble, volatile, poorly adsorbed, and rapidly degraded. Typical non-chlorinated solvents include xylene, acetone, ethyl benzene, and methyl isobutyl ketone. They are used as paint and varnish solvents, for cleaning machinery, in the production of rubber, in fuels and gasoline, and for other industrial uses. Generally, they are not as toxic as the chlorinated solvents, but they may irritate the eyes and skin and cause damage to the kidneys and liver.

Petroleum hydrocarbons.   Hydrocarbons are molecules consisting primarily of hydrogen and carbon. Those of lower molecular weight are generally soluble, volatile, rapidly degraded, and poorly adsorbed. As the molecular weight of the compounds increases, they become less volatile and less soluble. Typical petroleum hydrocarbons include natural gas (methane), aviation gasoline, fuel oil, kerosene, lubricating oils, and pitch and tar.

PCBs (poly-chlorinated biphenyls).   A family of synthetic chlorinated hydrocarbons in use since the early 1930s, these are generally insoluble, non-volatile, poorly degraded, and highly adsorbed. Because of their stability, they were widely used as electrical insulating fluids in transformers and capacitors, as well as in other specialty applications such as pressure-sensitive copying paper. Important proprietary names include Arochlor, Askarel, and Pyranol. PCBs are no longer manufactured, and their use is being phased out. PCBs accumulate in fatty tissues and have been shown to cause cancer in experimental animals, although some controversy exists over their effects on human health.

PAHs (polynuclear aromatic hydrocarbons).   More than 100 compounds exist in this family, including benzo-a-pyrene, acenaphthene, fluoanthene, and naphthalene. Their properties depend on the molecular weight of the compound. They are a common air pollutant when dispersed by the combustion of hydrocarbons such as coal, wood, and oil. Some PAHs have been shown to be carcinogenic in animal tests; they may build up in fatty tissues.

Pesticides.   These are generally insoluble, non-volatile, poorly degraded, and highly adsorbed. "Pesticide" is a general term for chemicals that are targeted at insects (insecticides), plants (herbicides), molds and mildews (fungicides), bacteria (bactericides), mites and ticks (acaracides), or rats and mice (rodenticides). Pesticides are very diverse, and more than 700 active ingredients (the actual chemical that does the killing) are registered in the United States. The most widely used fall into four groups:

1. Organochlorines: organic compounds containing chlorine, such as DDT, chlordane, Kepone, lindane, aldrin/dieldrin, methoxychlor, Mirex, and heptachlor. These have an environmental persistence of 2 to 15 years.

DDT (dichloro-diphenyl-trichloroethane) was one of the most widely used insecticides in the world because it is toxic to insects, but much less toxic to other animals; it is very persistent, continuing to exert its properties for a long time; and it is relatively inexpensive. Its degradation products are DDD (dichloro-diphenyl-dichloroethane) which has some toxicity to insects with lower toxicity to fish than DDT, and DDE (dichloro-diphenyl-ethane), with low toxicity to insects. Commercial DDT is a mixture of all three, with DDT predominating. DDT was very beneficial in the years after its discovery: it helped prevent epidemics of typhus in post-World War II Europe, and it formed the foundation for successful programs to control malaria and other insect-borne diseases in Africa and the Far East. It has now been superseded by less persistent insecticides in the developed world, but is still widely used in developing countries (Clark, 1989a).

The group of interrelated insecticides including aldrin, dieldrin, endrin, and heptachlor is extremely persistent, and the degradation products are also toxic. For example, heptachlor degrades to heptachlor epoxide, which is even more toxic than its parent; aldrin degrades to dieldrin. Because these chemicals are toxic to mammals as well as very persistent, they were largely withdrawn in the 1970s. In spite of this, they are widespread in the environment and continue to leach out of agricultural land into watercourses and oceans (Clark, 1989a).

2. Organophosphates: organic compounds containing phosphates, such as parathion, malathion, diazinon, TEPP, and mevingphos, with a persistence of a few weeks to a few years,

3. Carbamates: organic compounds containing nitrogen and oxygen, such as aldicarb, propoxur, Maneb, and Zineb, with a persistence of days and weeks,

4. Botanicals: natural insecticides generated by plants, such as rotenone and pyrethrum, with a persistence of days to weeks.

Pesticides—Jekyll-and-Hyde Substances

Pesticides have a Dr. Jekyll and Mr. Hyde personality. On the one hand, they greatly increase crop yields, eradicate pesky bugs, and enhance our quality of life. On the other hand, they are insidious and pervasive chemicals that have caused irreparable environmental damage (Carson, 1962). Pesticides are ubiquitous—their residues have been found in virtually all foods, plants, and animals.

→

Pesticides—Jekyll-and-Hyde Substances (continued)

Since 1939, when it was discovered that DDT is a potent insecticide, the manufacture of synthetic organic pesticides has flourished, and developing new ones has become an active industry. Worldwide, about 2.3 million tonnes of pesticides are used annually. In the United States, about 700 biologically active compounds are used to make some 50,000 pesticide products. Of these, herbicides account for 85%, insecticides 10%, and fungicides about 5%.

About 20% of all pesticides in North America are applied to lawns, gardens, parks, and golf courses. Since the mid-1970s, the use of slowly degradable chlorinated hydrocarbons has been replaced by more rapidly degradable pesticides, especially organophosphates and carbamates.

The biological effects of halogenated hydrocarbons—pesticides, solvents, PCBs, and so on—are very complex, and many aspects are still not well understood. First, they embrace an enormous variety of related compounds. Some families, like dioxins, PAHs, or PCBs, have many dozens of members which differ only slightly from one another chemically yet can have significantly different biological effects. Second, the analytical means of detecting them and separating the different members of the families have become available only in recent decades; for example, it was not until 1966 that it was possible to identify PCBs analytically—more than 30 years after they came into use. And it was not until several years later that PCBs could be separated from DDT. As late as 1980, toxaphenes, a widespread contaminant, could not be reliably detected (Clark, 1989a). Scientists have only recently had the tools to focus on these chemicals, and the data bank of knowledge is still small. Even today, despite vast improvements in electronics and laboratory analytical techniques, many organochlorines are probably being overlooked. To complicate things even more, the effects of these chemicals on different species can also differ significantly. Thus, it is understandable that considerable controversy exists about the effects of chemicals such as PCBs or dioxins on human health and the environment.

Inorganic Compounds

Inorganic compounds are defined by exception; that is, they are those chemical compounds that are not organic. Inorganic elements can be classified into three types:

- Major inorganic elements with their compounds, such as the silicates, sulphates, cyanides, or iron.
- Trace metals, such as aluminum, cadmium, copper, lead, mercury, nickel, selenium, zinc, and their compounds
- Arsenic and its compounds

The major inorganic elements and their compounds comprise the bulk of our Earth. These constituents include calcium, silicates, sulphate, and iron. These major elements and their compounds control the acidity and electrochemical potential of soil and water and thus control the action of the trace metals in the environment.

The trace metals are those elements that are present in trace, or very small, concentrations in geologic materials. The behavior of these elements is controlled by the chemistry of the major inorganic elements. A subset of this class comprises heavy metals, such as cadmium, lead, and mercury, which have densities that are at least five times greater than water.

In low concentrations trace metals can be essential to good health, but in higher doses they can be toxic. Molybdenum, for example, is required by the human body for certain enzymes to function properly, but too much will cause gout. A deficiency of iron can cause a blood disorder, anemia. On the other hand, there are some metals—such as lead—that have no known beneficial functions in the body, and even small exposures can be harmful, eventually causing damage to the nervous and reproductive systems and the brain.

An important property of metals is that, unlike organic compounds, they never degrade or decompose. Although they can be buried in landfills, they will eventually be remobilized and may then become a threat to the environment.

Microbiological Waste

This class of waste, also known as biomedical waste, contains toxic materials associated with living cells. The main categories we are concerned with are bacteria and viruses. Bacteria are single-celled microorganisms lacking chlorophyll; some of them perform useful functions in the human body. Diseases such as diptheria, tetanus, and botulism are caused by toxin-producing bacteria. Viruses are a group of infectious agents that are much smaller than bacteria and need to grow in an animal, plant, or bacterial cell. Some viral infectious diseases include smallpox, measles, chicken pox, the common cold, rabies, and viral pneumonia.

Microbiological waste can include human anatomical waste, consisting of human tissues, organs, and body parts; animal waste, including tissues, organs, body parts, carcasses, bedding, fluid blood, blood products, and items contaminated with blood; non-anatomical waste, including cultures, stocks, specimens submitted for microbiological analysis, human blood, or items contaminated with human blood; and contaminated sharps, including needles, blades, and glass or other materials capable of causing punctures or cuts. Microbiological wastes can come from hospitals, health-care centers, veterinary clinics, funeral homes, medical laboratories, medical research centers, and blood banks, as well as from private homes. Although these wastes should be incinerated or disposed of at a hazardous waste disposal facility, they sometimes wind up in municipal landfills.

Waste Characteristics

Some characteristics are more important than others in determining how wastes should be managed and disposed. The most important characteristics are discussed below.

Waste Volumes—A Key to Management

One key parameter is the size or volume of waste that must be managed. Industrial and municipal wastes are the most abundant. Microbiological and radioactive wastes are much smaller in volume.

The large volumes of municipal waste severely limit the options that are available for its safe disposal. For example, given the colossal accumulation of wastes at the Fresh Kills landfill in New York City, there are few alternatives to heaping it into a large pile. Virtually any other disposal or treatment method, except incineration, would be prohibitively expensive.

When waste volumes are kept small, however, the wastes become "manageable." Then some of the following advantages may be realized:

Treatment of the wastes via encapsulation or fixation in leach-resistant materials like cement may become economically feasible, thus rendering them more suitable for permanent disposal.

Because a smaller land area will be required for the disposal facility, more potential sites should be available for selection, which should result in a disposal site with better geological, groundwater, and other technical characteristics.

Different disposal options—such as burial deep underground or shipment to distant, technically superior sites—may become practical.

Smaller waste volumes means fewer disposal sites, thus allowing regional centers for disposal rather than a site at every municipality.

Most important, decreasing the volumes of waste also decreases the amount of harmful chemicals and materials that are placed into disposal facilities and eventually re-enter the environment.

In short, for wastes, small is beautiful. A cornerstone of any waste management program should be to minimize the quantity of waste. For this reason, the current programs for recycling, reducing, and reusing are critically important. In addition, all other avenues to reduced waste volume should be vigorously explored.

Thermal Value

An important statistic is that, before recycling, approximately 80% of municipal waste consists of organic materials. Even after an aggressive recycling program that includes composting, waste will consist predominantly of organic materials, although the fraction might be lower—perhaps 70%. This is true because both inorganics (metals, glass) and organics (paper, plastics, yard wastes) are removed by the recycling process in approximately the same proportions.

Even with a recycling rate as high as 50%, the residual waste will still contain a significant fraction of organic materials. Thus, residual waste has a high thermal value and can be used as fuel for incinerators. In other words, there is no technical reason why the municipal waste stream cannot support both a strong recycling program and incineration.

To assess the practicality of waste incineration, it is necessary to know the thermal content of waste. Some thermal values—the energy that can be obtained by burning a unit of weight—are listed below for different waste materials (Neal and Schubel, 1987):

Newspaper	13,880 kJ/kg (5,980 BTU/pound)
Books, magazines	13,490 kJ/kg (5,810 BTU/pound)
Cardboard	13,000 kJ/kg (5,600 BTU/pound)
Wood	19,969 kJ/kg (8,600 BTU/pound)
Coal	14,620 to 33,650 kJ/kg (6,300 to 14,500 BTU/pound)
Gasoline	48,750 kJ/kg (21,000 BTU/pound)

The Hazard of Wastes

Probably the most important factor in determining how a waste should be managed is its "hazard" or toxicity. For example, malathion and parathion are two closely related organophosphate pesticides which produce similar symptoms of poisoning in humans and other mammals. Should waste streams containing these substances be treated equally? The answer, in fact, is no. Parathion is considerably more potent than malathion, and so it should be managed and disposed of more carefully.

Table 6.1 provides a brief overview of the effects of different toxic compounds on the human body.

It would be convenient to have a simple index—say, from one to ten—to describe the hazard or toxicity of specific waste compounds. Unfortunately, life is not that simple. Developing toxicity levels for substances is difficult, costly, and controversial. There are many chemicals, and they have different effects on the various organs of the body. Thus, comparing the toxic effects of different waste compounds is like comparing apples and oranges.

Another complication is that some chemicals have an immediate toxic effect—that is, the effects are felt shortly after exposure—whereas other chemicals have a delayed toxic effect, not apparent until much later. For example, cancer may not appear until many years after a person's exposure to carcinogenic chemicals.

The timing of exposure is also important. An acute exposure or dose, delivered over a short period of time, may have different consequences than the same dose delivered over a long period of time,

Table 6.1.   Toxic compounds and their target organs (Griffin, 1988).

Liver-attacking (hepatotoxic) compounds
  carbon tetrachloride, tetrachloroethane
Kidney-attacking (nephrotoxic) compounds
  halogenated hydrocarbons
Blood-attacking (hematopoietic) compounds
  nitrobenzene, aniline, phenols, benzene
Nervous system-attacking (neurotoxic) compounds
  methanol, metals (manganese, mercury), organometallics (methyl mercury)
Consciousness-attacking (anesthetic/narcotic) compounds
  olefins, acetylene hydrocarbons, paraffin hydrocarbons, aliphatic ketones, esters, nitrous oxide
Cancer-causing (carcinogenic) compounds
  asbestos, arsenic, radioactive elements, nickel fumes, cigarette smoke, saccharin, coal tar

---

called a chronic exposure. For example, one worker might receive an acute exposure by breathing fumes from an accidental spill of a toxic chemical; a second worker, who works near the vats where this chemical is produced, might inhale the chemical in much lower concentrations under ordinary circumstances, but over several years. Although the total quantity of chemical inhaled by both workers might be the same, the effects of the acute and chronic exposure could be quite different, depending on the chemical.

Research is insufficient to understand the exact effects of many compounds. Typically, toxicological research has been conducted using animals and at doses much higher than would occur in any waste management scenario. Even so, testing a single compound for toxicity can take two to five years and cost as much as $1 million. To complicate matters, responses to toxics differ among animal species; for example, aspirin causes birth defects in rabbits but is considered safe for human use.

The method of exposure can have a dramatic effect on the potency of toxic compounds. For example, nickel can cause cancer if inhaled as a fume but not if ingested.

Epidemiological studies (the use of statistical methods to study diseases in large groups of people) are used to investigate the patterns of certain diseases, particularly those with delayed effects such as cancer. Typically, people exposed to a particular toxic chemical from an industrial accident, or people working under a high occupational exposure level, are compared with similar groups who are not exposed to see if there are statistically significant differences in their health. Usually these groups must be tracked for many years to gain relevant information, and even then it is difficult to say with certainty whether an observed effect was caused by exposure to a particular toxic compound or by some other stimulus.

In spite of these problems, many laboratory experiments and epidemiological studies have been performed to determine the degree and type of hazard associated with potentially toxic substances. Government regulatory agencies such as the U.S. Environmental Protection Agency (EPA) and Environment Canada in Canada, as well as international organizations such as the U.N. World Health Organization, evaluate the results of these investigations and develop regulations and guidelines to limit human exposure to specific toxic substances. These regulations and guidelines can form a means for classifying substances according to their toxicity.

An example of such a classification system is the one developed by the EPA for acute toxicity (U.S. Forest Service, 1984). The system is based on the lethal dose (LD_{50}) it takes to kill 50% of the organisms in a laboratory test. LD_{50} coefficients have been developed for a wide variety of substances. The EPA guidelines for acute toxicity are shown in Table 6.2.

Four categories of toxicity are defined, ranging from I (Danger Poison) to IV (caution). Each of these is defined for each of the three main means by which a human can be exposed to a toxic compound: by eating (ingestion or oral intake), skin (dermal) contact, and breathing (inhalation). Under inhalation, both particulates (dust) and gas are included. For many substances, toxicity can differ significantly depending on the route of entry into the body. Some toxics, for example, are particularly well absorbed from the digestive tract but not from the lungs or skin. In such cases, the oral dose would be the critical dose.

The most highly toxic substances are those that cause death or severe illness in very small doses (category I). These must be labeled DANGER POISON. Most household cleaning products are in category III and bear the word CAUTION on their labels. Consumers are advised to handle even category IV chemicals with caution, since these too can be poisonous in very large doses.

In addition, the EPA has adopted a classification system for carcinogens (cancer-causing substances). The system, shown in Table 6.3, is similar to that developed by the International Agency for Research on Cancer.

As we can see, the classification of substances by toxicity is a very complex topic. There are many variables involved, information on many chemicals is incomplete, and it is not possible to compare directly the hazards of acute versus chronic exposures, or carcinogenic versus non-carcinogenic substances. Nevertheless, toxicity is an exceedingly important characteristic and must be taken

Table 6.2 EPA guidelines for toxicity (U.S. Forest Service, 1984).

Toxicity Category	Oral LD_{50} (mg/kg)	Dermal LD_{50} (mg/kg)	Inhalation LC_{50} (mg/L dust)	(ppm vapour/gas)
I DANGER POISON	≤50	≤200	≤2	≤200
II WARNING	50–500	200–2,000	2–20	200–2,000
III CAUTION	500–5,000	2,000–20,000	20–200	2,000–20,000
IV caution	>5,000	>20,000	>200	>20,000

Note: mg/kg is the number of mg of the substance taken per kg of body weight. LD_{50} and LC_{50} are the lethal dose and lethal concentration, respectively, at which 50% of ingesting organisms die.

Table 6.3   EPA classification system for carcinogens.

Group A. Human carcinogen
There is sufficient evidence from epidemiological studies to support a cause-effect relationship between the substance and cancer.

Group B. Probable human carcinogen
B[1]: There is sufficient evidence for carcinogenity from animal studies and limited evidence from epidemiological studies.
B[2]: There is sufficient evidence for carcinogenity from animal studies, but the epidemiological data are inadequate or nonexistent.

Group C. Possible human carcinogen
There is limited evidence of carcinogenity from animal studies and no epidemiological data.

Group D. Not classifiable as to human carcinogenity
Data are inadequate or completely lacking, so no assessment as to the substance's cancer-causing hazard is possible.

Group E. Evidence of noncarcinogenity for humans
Substances in this category have tested negative in at least two adequate (as defined by the EPA) animal cancer tests in different species and in adequate epidemiological and animal studies. Classification in group E is based on available evidence; substances may prove to be carcinogenic under certain conditions.

into consideration when making waste management decisions. The development of a simplified "toxicity index" that could be applied to all waste types would be very useful.

Risk Assessment

To understand the risk or hazard associated with a disposal or waste management facility, it is necessary to conduct a detailed risk assessment or pathways analysis. Over the past decade, risk assessment methodology has evolved considerably, and there is now a general consensus on the approaches and methods that should be applied. Risk-based management allows decisions to be made regarding waste facilities or contaminated sites that are quantitative, objective, and consistent for different sites and facilities with different characteristics. Needless to say, risk assessment is complex and involves many scientific disciplines.

Figure 6.3 shows the pathways by which contamination might escape from a conceptual waste disposal facility and travel by groundwater, surface water, and/or air to reach the environment and give a dose to receptors such as humans, animals, and plants.

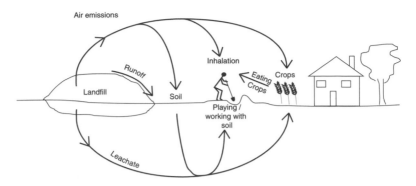

6.3   Schematic of risk assessment pathways.

Risk assessment begins by defining the source term—that is, the volumes and toxicity of the materials in the facility. Then computer simulations track the contaminants as they escape or are emitted from the facility and travel through air, surface, and subsurface routes to reach humans. This analysis requires that all possible pathways by which contaminants can travel be identified and quantified. For example, for a landfill, the process by which a subsurface liner might degrade and allow waste compounds to leach into the groundwater would be modeled. The pathway that the contaminants travel through the groundwater to reach water wells and plants would be modeled, as well as the amount of the contaminant that then gets ingested by a human using the well and eating the plants. Similarly, all other realistic routes would be included. Doses are calculated for a hypothetical individual or some critical population group residing in a location where exposures are highest. The results are usually stated in relative terms in chances per million of contracting, say, cancer, after a lifetime of exposure (70 years) to the emissions from the facility.

This is a complicated process and requires a knowledge not only of the toxicity of wastes but also of their physical and chemical forms. In addition, detailed knowledge of the natural setting of the waste disposal facility is necessary, including its geography, geology, surface and groundwater, vegetation, human settlement, and so on. Risk assessment is becoming the preferred method for making decisions regarding waste management. For example, risk assessment calculations are performed to justify environmental regulations and to design and license waste management facilities.

A risk assessment is typically used in the following manner in designing a waste disposal facility. The source material and its toxicity are calculated, based on data from waste generators. Then a disposal facility is designed to contain this waste for a selected site. A risk assessment is performed using this information and yields doses to critical population groups in the vicinity. If it turns out that the doses are unacceptably high, the design of the waste disposal facility can be adjusted—for example, by increasing the thickness of the engineered barriers preventing escape of the wastes until the calculated doses to the critical population group are acceptable. In this manner, the future performance of a disposal facility can be evaluated, allowing weaknesses to be identified and modifications to be made as necessary, before the start of construction. These calculations can also be used as part of the licensing process to demonstrate to the regulatory agency that the facility will be safe.

Hazardous Lifetime

Many wastes decompose with time. They degrade into more stable forms, and their toxicity usually decreases as time passes. Thus, an important waste characteristic is the length of time over which a substance remains hazardous. An understanding of this characteristic is essential to managing wastes properly. For example, if a particular waste stream degrades to an innocuous level in a decade or two, it would be a waste of resources, although perfectly safe, to build a disposal facility to last for centuries. In a case that more closely reflects our current practice, when a waste stream has a hazardous lifetime spanning several centuries, it would be unsafe and irresponsible to construct a disposal facility that only provides security for decades.

The decomposition of many organic compounds is exponential with time, often obeying first-order rate constants. That is, their decomposition can be described in the same way as radioactive decay, using the concept of half-life—the period of time in which the compound degrades to half its original level of activity. These half-lives may be years, decades, or even centuries, particularly for compounds such as PCBs. For example, the half-lives of toluene and dieldrin are 10.4 hours and 2.96 years, respectively. The hazardous lifetime of a specific compound, then, can be measured in the number of half-lives required for it to decompose to an acceptably low concentration.

The specific duration required for organic compounds to decompose to less complex forms, and ultimately to water and carbon dioxide, depends on many factors, such as the availability of oxygen (aerobic or anaerobic conditions), the chemical composition of the wastes, the presence of water or leachate, temperature, and the permeability of the waste mass. Most of the decomposition of organic wastes in a landfill occurs under anaerobic conditions (lacking oxygen) and results in the production of methane and carbon dioxide. This process depends on site-specific conditions, but it takes approximately 50 to 150 years.

In contrast, inorganic components of waste, such as heavy metals and their compounds, never degrade or decompose. Thus, they effectively have an infinitely long hazardous lifetime. Under certain chemical conditions, however, these inorganic compounds can be adsorbed, or bound up, with other wastes and landfill contents for very long periods until the chemical conditions that promote adsorption change. In other words, in a landfill, the hazard decreases with time down to a base level, *which is greater than zero*. The hazard or risk of a landfill never goes to zero owing to the presence of inorganic materials. This concept is illustrated schematically in Figure 6.4.

This very important waste characteristic of longevity is not generally taken into account in the design and regulation of municipal or industrial landfills. Most landfills are only required to be monitored for approximately 25 years after closure. This period is inadequate even for the organic or short-lived component of waste, and it is woefully inadequate for the inorganic component.

In summary, municipal, hazardous, and industrial wastes con-

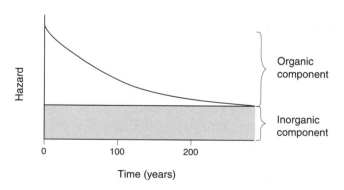

6.4   Landfill hazard as a function of time.

tain both organic and inorganic constituents. The former have a relatively short hazardous lifetime (a few to a few hundred years), whereas the latter have a long lifetime (effectively forever).

Radioactive wastes share this dual character. Nuclear wastes consist of two types of material, fission products and transuranic elements. Fission products decay to stable configurations in periods from a few minutes to a few hundred years, depending on the particular nuclear isotope (there are many different fission products created in the nuclear fuel while it is in a nuclear reactor). Transuranic elements, on the other hand, decay to stable configurations much more slowly. There are many transuranics formed in nuclear reactions, and some of these take thousands or even tens of thousands of years to reach a stable form. During the period when the radioactive elements are decaying, they emit radiation and therefore are hazardous. The transuranic elements are, for practical purposes, hazardous forever: compared to the short span of a human lifetime, tens of thousands of years can be considered as an infinite time.

Although radioactive wastes are seen by the public as a menacing, highly toxic, and largely incomprehensible threat, their health effects are similar in character to those of other waste types. As shown earlier in this chapter, the chief impact of radioactive wastes is that they cause cancer. Thousands of non-radioactive wastes also cause cancer. Furthermore, the hazardous lifetimes of radioactive wastes are comparable to those of the other, more common waste types. The relatively short-lived fission products are analogous to the organics, and the long-lived transuranics are similar to inorganics such as heavy metals.

Two important points emerge from this discussion. First, all waste classes should be managed in a manner that takes into account their hazardous lifetime. In particular, disposal facilities should be designed so that they provide protection to the environment for both the short term and the long term. The latter timeframe has generally not been addressed.

Second, the common perception that radioactive wastes are very different from other, more common waste types is largely a misconception: radioactive wastes share many important characteristics with other wastes. This suggests that we would benefit from a uniform and standardized approach to all wastes. This has many implications. First, there is a need to develop disposal facilities that provide protection over the long term. In this regard, disposal methodologies proposed for nuclear wastes are more advanced than those for other waste types. The fundamental similarities between waste types suggest that a common regulatory scheme should be used for all wastes, rather than having different agencies issuing different regulations.

Discussion Topics and Assignments

1. List hazardous materials that are discarded in garbage from households and businesses. List hazardous wastes that arise in a university.
2. Find out how much municipal solid waste your community generates and determine its percentage composition, using the categories shown in Figure 6.1. How do they compare? What are the reasons for any differences?
3. Assume that the legal limit for disposal into landfill of benzene in sludge from a factory is 5 ppb (parts per billion). Fifty days after disposal of some sludge, a sample was analyzed and yielded a concentration of 1.35 ppb. If the first-order removal rate constant for benzene is 0.00345/hr, what was the concentration of benzene when the sludge was brought to the landfill? Should the disposal have been allowed? [$C(t) = C(0)*exp(-kt)$ where k is the first-order removal rate constant, and $C(t)$ is the concentration at time t]
4. Contact the EPA or other environmental and health organizations and find examples of chemicals that fit each of the categories in tables 6.2 (toxicity categories) and 6.3 (carcinogen classification).
5. Draw up a list of information that you would require to conduct a risk assessment of a municipal landfill serving a community of 400,000. Qualitatively, describe the steps in the assessment and which information is required in each step.

Suggested Reading

Carson, Rachel. 1962. *Silent Spring*. Cambridge, Mass.: Riverside Press.

Fletcher, W. W. 1974. *The Pest War*. Oxford: Basil Blackwell.

Glenn, W. M. 1989. *A Reporter's Guide to the Chemicals in Today's Headlines*. Toronto: Ontario Waste Management Corporation.

Griffin, R. D. 1988. *Principles of Hazardous Materials Management*. Chelsea, Mich.: Lewis.

Harte, J., C. Holdren, R. Schneider, and C. Shirley. 1991. *Toxics A to Z: A Guide to Everyday Pollution Hazards*. Berkeley, Calif.: University of California Press.

Klaassen, C. D., M. O. Amdur, and J. Doull. 1986. *Casarett and Doull's Toxicology: The Basic Science of Poisons.* 3rd edn. New York: Macmillan.

Parmeggiani, L. (ed.). 1983. *Encyclopaedia of Occupational Health and Safety.* 3rd edn. Geneva: International Labour Office.

Paustenbach, D. J. 1989. *The Risk Assessment of Environmental Hazards: A Textbook of Case Studies.* Somerset, N.J.: John Wiley & Sons.

Rathje, W. L., and L. Psihoyos. 1991. Once and Future Landfills. *National Geographic,* May, pp. 116–134.

Reynolds, A. B. 1996. *Bluebells and Nuclear Energy.* Madison, Wisc.: Cogito Books.

LANDFILLS
How Do They Work?

Throughout history the landfill has been the bottom line of waste management: this is where the buck stops. Because of its fundamental importance, a thorough understanding of how a landfill is designed and operated is necessary.

Municipal landfills are the most common; they accept our household garbage and often take some commercial and industrial waste as well. They are generally operated by local municipalities, although some are privately owned. We are all stakeholders in the landfill, however, contributing our share when we place our garbage by the curbside each week.

Another type of landfill is that used by industries. These are generally situated near the industrial plants they serve and are usually dedicated to the specific process wastes produced at the factory. In terms of design, they do not differ significantly from municipal landfills. There are also "secure" landfills for hazardous waste, known in the United States as Subtitle C hazardous waste disposal sites.

These different landfills do not differ greatly in design, and the following descriptions are largely applicable to all of them. A number of specific case histories are presented in chapter 11.

Siting

Historically, it has been convenient to site landfills in depressions such as ravines, canyons, abandoned quarries, and open pits that could be easily filled up. These locations were selected largely on the basis of convenience, with proximity to the centers being served and price of land being key parameters. In the early 1900s, and even as late as the early 1970s, landfills were seen as an ideal way of "reclaiming" swamps and wetlands. With the loss of natural areas to urbanization and agriculture, and with the recognition that wetlands play an important role in the ecosystem, this practice is no longer condoned.

It is now recognized that one of the most effective ways of protecting the environment is by carefully siting landfills at locations that provide natural security. In particular, the geological formation should contain or naturally attenuate contaminants to acceptable concentrations. If natural attenuation is not possible, then engineered barriers must be incorporated. Today, the trend is toward both incorporating extensive engineered barriers and seeking sites that offer geological containment.

Landfill site selection is a complex process in which many technical factors must be considered: surface and ground water, the presence of suitable soil and natural conditions, transportation routes, topography, the presence of endangered or sensitive species, and much more. In addition, land planning and politics play an enormous role, with the latter often the dominating force. Site selection in our modern era of NIMBY, legal obstruction, and citizens' protest movements, could easily be the subject of many books. Only some basic technical requirements of siting are discussed here; the NIMBY phenomenon is described in chapter 12.

Site selection often takes place in two phases. The first looks at a regional area, using available information such as geological maps and reports on meteorology, flora, fauna, and other features. On-site investigations at this phase are generally restricted to visual inspections. The regional phase concludes with the identification of several specific areas that might be suitable for a landfill. The second phase focuses on these specific areas and includes detailed on-site studies such as soil drilling and sampling, groundwater investigations, and weather monitoring. In this phase, the facility design is integrated with the site so that the natural setting and facility features combine to offer maximum containment of the wastes and protection of the environment.

A well-defined process must be established so that decisions will be made in a systematic and documented manner as the siting progresses from large regions to specific areas. This is critical to ensuring that the final site is defensible, acceptable to the public, and likely to gain the required regulatory approvals.

Traditionally, regional siting has involved a technique called
"constraint mapping." This involves defining characteristics that are
undesirable for landfill sites, then systematically eliminating areas
that possess those characteristics. Usually, computer methods such
as Geographic Information Systems are used to produce the maps.
Input from the public and regulators is important in defining the
constraints to be applied. Typical constraints might be avoiding
floodplains, maintaining at least 100 meters distance from lakes or
streams, leaving a buffer zone of at least 400 meters around the nests
of endangered bird species, or avoiding archeological and historic
sites. Overlay maps can be produced on transparencies, in which
each overlay has a different constraint shaded. When the overlays
are superimposed, as shown schematically in Figure 7.1, the areas
that show through clearly are those that have no constraints; these
can be investigated further as potential landfill sites.

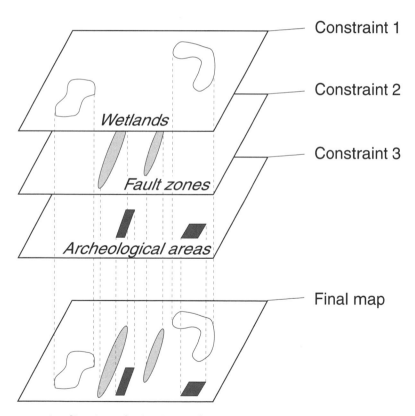

7.1   Application of criteria overlays in constraint mapping for site
selection.

Positive attributes can also be defined at a relatively early stage of siting. These might include proximity to transportation, availability of electricity and other utilities, impermeable subsoils, or a deep water table.

These constraints and attributes are systematically applied in a predetermined manner until the array of sites under consideration has been reduced to one or a few preferred areas.

In local siting, a number of detailed studies are undertaken on the preferred area(s) to determine:

- Geologic setting including soils, bedrock, and their spatial variations: Areas should be avoided that have karst landscape (limestone bedrock with dissolution caverns and potholes), unstable slopes prone to landslides, subsurface mines, or seismic activity.
- Surface water and groundwater: A thorough understanding of the latter is essential, including groundwater quality, flow directions, location and extent of aquifers, and location of wells. Areas to be avoided include floodplains and wetlands.
- Biological resources and uses. Areas to be avoided include those with endangered species.
- Local microclimates and dispersion patterns.
- Adjacent land use: Landfills should not be sited near airports because they attract birds, which can cause airplane crashes.
- Social and cultural patterns.

To aid in the comparison of sites, a numerical ranking can be used by developing a set of N criteria and then assigning each one a weighting factor, W_i. Typical criteria include subsoil permeability, soil cation exchange capacity, proximity to transportation, proximity to streams or lakes, seismic activity of the area, steepness of topography, depth to the water table, and proximity to groundwater wells. The ranking score R_j for site j is then:

$$R_j = W_i S_{ij} + W_z S_{zj} + \cdots + W_n S_{nj}$$

S_{ij} is the criterion score with regard to criterion i for site j. For example, a site that is considered good with respect to criterion i could be given a criterion score of 3; a site that is moderate would receive a 2; a site that is fair, a 1; and one that is poor would receive a 0. The W_i values stay the same, whereas the S_{ij} values change for each site. In this way, different sites can be compared in an objective, quantitative fashion.

An important part of both regional and local siting is integrating the investigations and the siting process with an information program that seeks the input of local people and politicians and keeps them fully aware of progress. Typically, community meetings are used to gain consensus on which constraints are to be applied at the regional level, and what criteria and weighting factors should be used in comparing and selecting specific sites.

Landfill Design: Anatomy of a Landfill

This section describes the anatomy of landfills and how they are constructed and operated. Typically, there is little difference between municipal and industrial landfills, so the description that follows is, in most regards, typical of both.

Basic Shape

The basic method of landfilling has not changed dramatically over the years. Today, natural depressions are still used where possible. Where they are not present, shallow holes are dug into the soil, using bulldozers and other excavating machinery. Landfills are generally designed in a number of cells; within each cell the garbage is placed in a series of "lifts" (the waste emplaced in one day) to allow the landfill to grow in the vertical as well as the horizontal direction. The patterns of cell and lift construction are carefully planned for access, to minimize travel distances and to allow travel over previously filled areas so they will be compacted. A major advantage of landfills is that little upfront capital is required because the cells are developed as required.

There are two primary methods of developing a landfill—the trench and area methods—although variations and combinations are possible (O'Leary et al., 1986). In the former, a trench is initially excavated, and then solid waste is spread and compacted in it in lifts. The soil removed from the trench is used as cover material. In general, the trench method is used in relatively flat terrain where the groundwater table is relatively deep and the soil is more than two meters thick.

The area method, which is suitable for more varied topography, is used when large quantities of waste must be disposed of. Solid waste is spread and compacted on the natural surface of the ground, and cover material is brought from off-site.

In recent times, with the difficulty of siting new landfills, it has become necessary to raise existing landfills to significant heights. In such circumstances the trench method is used at the base, and the area method is then used to build the landfill up to its final height.

A considerable amount of engineering planning and design is necessary for a modern landfill. Projections of future waste generation rates must be made, and based on these, the landfill is designed. Layouts and cross-sections are drawn that allow for the projected growth in a rational manner. Site drainage, roads for vehicle access, and support facilities such as buildings and retention ponds, are included in the design.

By the 1970s and 1980s there was increased emphasis on the use of technology and engineering for landfills. The main features added to landfills over the past few decades include the following:

- Daily covering of the garbage with soil or other inert material
- Construction of an impermeable liner below the landfill
- Placing an impermeable cover over the landfill
- Collection of leachate
- Collection and extraction of landfill gases
- Monitoring the landfill and its environment

Figure 7.2 shows, in schematic form, the general layout and cross-section of a modern landfill. Figure 7.3 shows an aerial view of a completed landfill cell next to a new cell that has just had a geomembrane bottom liner installed.

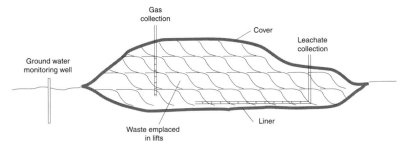

7.2   Schematic cross-section of a municipal solid waste landfill.

7.3   Aerial view of Trail Road landfill, Ottawa, Ontario, showing
a completed cell with composting windrows, a new cell with
geomembrane bottom liner, and a gas flaring stack (courtesy of
Regional Municipality of Ottawa Carleton).

Protective Cover

In terms of environmental protection, the final cover, or cap, over a
landfill is a key part of its overall design. Its primary objective is to
isolate the interior of the landfill from the infiltration of water and
thus to prevent the generation of leachate. The cap should be less
permeable than the bottom liner to prevent a build-up of water inside the landfill. The cover should be designed to promote the
growth of vegetation in order to protect the landfill from erosion
and intrusion by humans, burrowing animals, and plant roots, and
also to improve the esthetic appearance.

A typical final cover for a landfill can be from 0.6 to 2 meters thick.
An example of a final cover recommended by the USEPA is shown
in cross-section in Figure 7.4. Clay layers and very low-permeability
plastic liners (geomembranes), such as high-density polyethylene
(HDPE), are used either alone or together to prevent the entry of water.
The drainage layer diverts any infiltrating water to a collection and
removal system. Other components may consist of, in various combinations, geotextiles, sand, coarse sand and gravels, topsoil, and

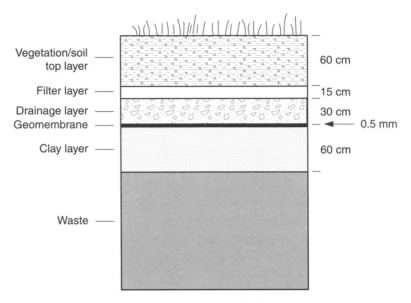

Vegetation/soil top layer	60 cm
Filter layer	15 cm
Drainage layer	30 cm
Geomembrane	◀— 0.5 mm
Clay layer	60 cm
Waste	

7.4   A landfill cover system recommended by USEPA.

grass. Geotextiles are essentially durable clothlike sheets that prevent the materials of one layer from mixing with the materials of the adjacent layer, usually to prevent a drainage layer from becoming clogged by finer particles. They allow gases and liquids to pass. If there is concern about future human intrusion, perhaps after the landfill has been abandoned, a layer of concrete or larger rocks (riprap) can also be included, although this is not commonly done.

The final surface is graded to promote runoff of rainwater, and vegetative cover is planted; this not only prevents erosion but also promotes runoff and evaporation. Surface-water flow usually starts as sheet flow, but it can concentrate over short distances to form rills and then gullies which erode the cap and carry sediments downward. With the modern trend to higher landfills, a heavy rainstorm can cause considerable damage. The choice of soils, topographic shape, and plants, combined with flow velocity and erosion analyses to minimize such erosion (for example, refer to Oweis & Khera, 1998), is an important part of landfill design. Overall site drainage needs to direct rainwater away from filled and operating areas, thus preventing it from infiltrating into the landfill.

A significant difference between covers and bottom liners is that the latter are constructed on solid foundations so that there will be very little settlement. Covers, however, will be subjected to considerable settlement as the wastes below gradually decompose and compact. Because of the landfill's heterogeneous contents, settlement will not be uniform, and stress on the cover will cause it to crack. In addition, freeze/thaw cycles during winter, spring downpours, and wind erosion will all act quietly but persistently to degrade the cover. For these reasons, not only must the cover be designed and constructed with extreme care: it will also be necessary to inspect and perform repairs and maintenance for many decades, if not centuries, after closure, until the landfill no longer poses a hazard.

Bottom Liners: Barriers against Leaking Leachate

The bottom liner, the layer that underlies the landfill, is the single most important element of a landfill. The purpose of this component is, first and foremost, to contain any leachate generated in the landfill and prevent it from leaving the site and contaminating nearby surface waters and groundwater. Landfill liners are composed of essentially the same elements as the final cover, although the order of their arrangement is somewhat different.

In 1982 the U.S. EPA banned reliance on clay liners alone for hazardous waste sites and specified the use of single or double liners made of impermeable synthetic membranes. Regulations issued by the EPA in 1991 require that new municipal landfills have a minimum of six layers of protection between the garbage and the underlying groundwater. At present, an estimated two-thirds of U.S. landfills do not have liners.

Figure 7.5 shows a cross-section of a state-of-the-art liner system installed at a modern landfill (ECDC Environmental, n.d.). The main barriers to leakage are 1.2 meters of remolded clay and a plastic liner. The latter is typically high-density polyethylene with a thickness of approximately 2 millimeters (i.e., about 80 mils; one mil is one thousandth of an inch). Figure 7.6 shows a bottom liner being installed.

Clay that has been remolded—reworked and compacted to make it very watertight—is a preferred liner for landfills. Because natural geologic materials have remained in place for many thousands of years, it can be expected that clay liners will remain undisturbed for similar periods. The most serious problem with clays is their tendency to crack when they become dry.

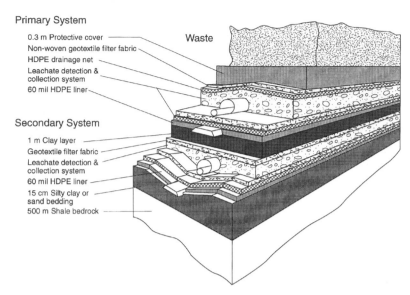

Primary System

- 0.3 m Protective cover
- Non-woven geotextile filter fabric
- HDPE drainage net
- Leachate detection & collection system
- 60 mil HDPE liner

Waste

Secondary System

- 1 m Clay layer
- Geotextile filter fabric
- Leachate detection & collection system
- 60 mil HDPE liner
- 15 cm Silty clay or sand bedding
- 500 m Shale bedrock

7.5   Detail of a bottom-liner system at East Carbon landfill, Utah (courtesy of ECDC Environmental).

7.6   Installation of bottom liner at Trail Road landfill, Ottawa, Ontario (courtesy of Regional Municipality of Ottawa Carleton).

Plastic liners, called "geomembranes," provide good containment in the short term, but they can easily be damaged by heavy equipment; moreover, we have no experience in assessing their resistance to degradation over long time spans such as a century or more. Clay and geomembranes used in the cover and liner systems are discussed in greater detail in chapter 10.

An important part of the liner system is a leachate collection system located above the impermeable layer. Horizontal perforated pipes are inserted in each drainage layer to collect leachate. The purpose is to reduce the head (the pressure) on the impermeable part of the liner system; the EPA requires that buildup of leachate head in the upper drainage layer of a composite liner system should not exceed 30.5 cm (one foot). The lower drainage layer is monitored to detect any leakage from above. Leachate is pumped to surface holding tanks and either treated on site or sent to a sewage treatment plant. Alternately, the leachate might be recirculated back into the landfill to promote decomposition. If the leachate is treated on site, the treatment plant should be designed to deal with a range of leachate compositions reflecting the changes that are expected over time.

A Different Kind of Liner

At the Regional Municipality of Halton, west of Toronto, Ontario, an innovative bottom liner system was used in a newly constructed landfill. The main protective bottom layer consists of clay, which is common practice. But instead of a synthetic (HDPE) membrane to provide a secondary bottom liner as is the norm, they use a hydraulic barrier created by pumping water into a 30-centimeter-thick layer of 5-centimeter stones below the clay liner at the bottom of the landfill. A positive pressure is maintained that prevents leachate from migrating downward into the underlying native soils (Lowry and Chan, 1994).

To provide safety through redundancy, the EPA requires that dual leachate systems as well as leak detection systems be installed as part of liner systems for new landfills in the United States.

Landfill construction requires that large rolls or panels of geomembrane liners be joined in the field. This critically important operation requires careful attention so that the seams do not form zones of weakness or leakage. The techniques used to join geomembranes are described in chapter 10.

Operation

Landfills are far from elegant. Most of us have made a visit to the local dump to drop off a load of tree cuttings, an old sofa, or other items too cumbersome to be picked up at curbside. As we enter the landfill site, our first impression is of dust, litter, and pandemonium. Trucks and cars are traveling in many directions, usually trailing plumes of dust. As we near the tipping face, the amount of wind-blown paper and other debris increases—as does the smell. Bull-dozers are moving here and there, shifting mounds of garbage in seemingly random directions and compacting the garbage below. Seagulls or crows congregate at the face where garbage is being placed, constantly wheeling, landing, and taking off again (Fig. 7.7); their plaintive screeching is a trademark of landfills.

Behind the apparent disorganization, however, is a complex and well-managed operation. In addition to the tipping of garbage, other important facilities and operations at a modern landfill can include:

- A scale house near the landfill entrance where incoming garbage trucks are weighed and their loads inspected.
- A depot where household hazardous wastes can be dropped off. These special wastes include paints, oils, aerosol containers, pesticides, batteries, propane tanks, antifreeze, and expired pharmaceuticals.

7.7   The plaintive screech of seagulls is a trademark of landfills.

- An area where trees and other yard debris can be dropped off: an industrial-sized shredder reduces branches and bushes to mulch, which can be used in the municipality's landscaping programs, sold, or used as cover material at the landfill.
- An area where leaves, grass cuttings, and other organic materials are placed in windrows for composting.
- An area for large recyclable materials such as "white goods" (household appliances such as refrigerators, stoves, washing machines, or lawnmowers) and tires. The white goods are periodically removed by scrap-metal dealers who recycle the steel, copper, and other useful materials. Tires are also removed periodically to be used for making rubber products or for incineration for energy.
- An administrative building contains offices, often a visitors' center with displays, and a laboratory for analyses in the extensive monitoring programs that form part of the operation of a modern landfill.

Most of the action takes place at the working face of the daily lift. Important tools in the operation of the landfill include bulldozers for spreading and compacting trash to the proper thickness and horizontal extent of the lift. Separate compactors are used to compress the garbage more and ensure that a maximum density is achieved, so that subsequent settlement is minimized. Figure 7.8 shows a compactor

7.8 A compactor at work (courtesy of Regional Municipality of Ottawa Carleton).

at work, and Figure 7.9 shows typical heavy equipment used at landfills. The size of the daily working face is kept to a minimum to conserve the amount of cover material and reduce windblown litter.

The amount of cover material used in a landfill can be quite substantial, and its supply and stockpiling is an important part of planning and operation. As a cell is filled, refuse is generally not left uncovered for more than 12 hours. At least 30 centimeters of cover is placed if an area will not be used again for a period of a few days or weeks; generally, at least 60 centimeters is placed once an area is completed.

The placement of daily cover prevents rodents from burrowing or tunneling in the waste; it keeps flies from emerging, and it discourages birds from scavenging. It minimizes moisture infiltration and thus the generation of leachate. It controls blowing paper and

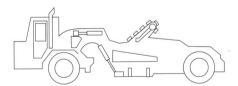

Compactor

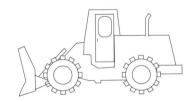

Wheel Tractor Scraper

Backhoe Loader

7.9   Typical equipment used at landfills

litter, which is an important esthetic consideration where residential or commercial areas are located nearby. Finally, it reduces the
potential for fires in the garbage.

An operating landfill usually includes fences erected specifically
to catch wind-blown litter. Often, a landfill also has an active program to minimize the number of birds, primarily seagulls. Rodent-
control programs are also an essential part of landfill operation.

Landfill Dynamics: Decomposition

A landfill is a dynamic entity which goes through a varied and evolving lifespan. Although the hustle and bustle that characterizes an
operating landfill is no longer evident once it has been closed and
capped, action continues below the cover. At this point, the primary
"living" mechanism is the decomposition of the organic material
in the waste.

Decomposition is the breakdown of the organic materials in the
landfill, and it is important that this process be understood. The
three main consequences of decomposition are:

- Generation of acid leachate
- Generation of potentially explosive landfill gas
- Settlement of the landfill, with potential cracking and deg-
  radation of the cover

There are many misconceptions about this process. On the one
hand, the landfill is seen as a source of rich, moist humus that can
be recycled to greenhouses, farmers' fields, and back to our home
gardens. At the opposite extreme, biodegradation in the landfill is
envisioned as the source of toxic leachate that can pollute our drinking water. In reality, very little has been known until recently about
the processes that occur inside landfills.

The Garbage Project conducted by the University of Arizona,
described in chapter 6, has unearthed a wealth of information about
the contents of landfills (Rathje & Psihoyos, 1991). They found that
although biodegradation takes place, it does so on a much smaller
scale than had previously been thought. The authors observed,
"Well-designed and managed landfills seem to be far more apt to
preserve their contents for posterity than to transform them into
humus or mulch. They are not vast composters; rather they are vast
mummifiers." This is so, at least in part, because low moisture content (generally less than 20% by weight; see Oweis & Khera, p. 295)

inhibits decomposition. Modern landfills minimize moisture content by the use of impermeable landfill covers which prevent the entry of water, as well as by leachate control systems which remove leachate from the bottoms of the landfills.

The Garbage Project noted that organics like food and yard wastes were the only items that could be considered truly vulnerable to biodegradation under normal landfill conditions. The decomposition process is highly complex and is exceedingly variable from place to place throughout the landfill. Some archeological digs have found that even after 2,000 years, much garbage has not fully decomposed.

Certain clues have hinted at this fact. For example, a landfill's methane generation in most cases amounts to less than half of what is theoretically achievable. Landfills tend to vent methane in relatively copious amounts, at least in terms of economic energy extraction, for a decade or two, and then production drops off relatively quickly.

Refuse typically consists of 40% to 50% cellulose, 12% hemicellulose, 10% to 15% lignin, and 4% protein on a dry weight basis, with the cellulose and hemicellulose accounting for 91% of the methane generation at a landfill (Barlaz et al., 1990). Rubber and most plastics are relatively resistant to biodegradation.

Biological decomposition in a landfill takes place in three stages, each of which has unique characteristics (McBean et al., 1995). These are described below and summarized in Table 7.1.

Aerobic Decomposition

This process requires the presence of oxygen and is the first to occur; it lasts until the available oxygen is consumed. This stage usually lasts for a short time, although it can continue in smaller, localized areas near the surface where a continued supply of air can be obtained. A general relationship for this process is:

Table 7.1.   The three stages of biodegradation in a landfill.

Stage	Products	Notes
Aerobic	CO_2 + Heat + H_2O	Needs O_2; short-lived
Acetogenic	CO_2 + H_2O + acids	Anaerobic; creates aggressive leachate
Anaerobic	CH_4 + CO_2 + H_2O	Long-lived; methane can be used as energy source

Biodegradable waste + O_2 → CO_2 + H_2O + heat + degraded waste

Note that no methane is produced in this stage. The aerobic microorganisms, however, produce carbon dioxide levels as high as 90%, and the temperature can rise to 70°C. The elevated level of carbon dioxide leads to the formation of carbonic acid, which results in acidic conditions in the waste leachate.

Acid-phase Anaerobic Decomposition

This is the second stage of waste decomposition. It involves different microorganisms that become dominant as oxygen is depleted and continue the decomposition process. Large quantities of organic acids, ammonia, hydrogen, and carbon dioxide are produced— hence the name "acid" or "acetogenic" phase for this stage. Acid fermentation prevails, yielding high levels of carbon dioxide, partially degraded organics (especially organic acids), and some heat. The process can be described as follows:

Biodegradable waste → CO_2 + H_2O + organic acids + degraded waste

Note that no methane is generated during this stage of decomposition. The production of carbon dioxide and large amounts of organic acids results in an acidic leachate, with the lowering of pH to about 5.5 to 6.5; this in turn causes other organics and inorganics to dissolve. The outcome is a chemically aggressive leachate with high specific conductance caused by the materials it has dissolved.

Anaerobic Decomposition

The third and final stage of biodegradation, also known as the methanogenic stage, is the longest-lasting and most prevalent. As the oxygen becomes depleted, anaerobic methanogenic bacteria become dominant. These organisms produce carbon dioxide, methane, water, and some heat. The gas methane is a particularly important product which is extracted at many landfills and used as an energy source. Typically, the organisms work slowly but efficiently over many years to decompose the remaining organic constituents of the waste. The methanogenic bacteria utilize the end products of the previous anaerobic acid stage—hydrogen and acetic acid (CH_3COOH)—as follows (McBean et al., 1995):

$$4H_2 + CO_2 \rightarrow CH_4 + 2H_2O$$

$$CH_3COOH \rightarrow CH_4 + CO_2$$

Consumption of the organic acids raises the pH of the leachate to near neutral—that is, a pH of 7 to 8. As a result, the leachate becomes less aggressive chemically and possesses a lower total organic strength. Some organic acids are converted directly into methane, while others go through intermediate steps.

The gases nitrogen (N_2) and toxic hydrogen sulfide (H_2S) may also be generated during anaerobic decomposition. Nitrogen is produced by a process called "denitrification" in which the nitrate ion is reduced. Hydrogen sulfide is produced by sulfate-reducing microorganisms. Hydrogen, which was produced in the acid-phase stage, is consumed during this stage and is generally not found in the presence of methane. Small amounts of many other gases can be generated, depending on the composition of the waste (see Table 7.3 below).

The methanogenic stage may not commence until six months to a few years after waste emplacement. The onset of this stage is accelerated by a high water content which causes anaerobic conditions. The optimal pH for the methanogenic phase is near neutral. The best temperature conditions are in the range of 30° to 35°C, with a second but lesser optimum around 45°C; most landfills are in the former temperature range.

It should be noted that oxygen is toxic to the methanogenic bacteria. In producing methane for energy, it is important that the extraction wells do not pump so hard that air is pulled in from the landfill surface, because this will poison the methane-generating species.

Working with Biodegradation

In summary, the biodegradation process breaks down complex organic molecules into simpler molecules. Although there are several intermediate steps, the final products are water, carbon dioxide, and methane (methane is the principal constituent of natural gas). It should be noted that the process of decomposition destroys some hazardous organic compounds—that is, it breaks them down into harmless components. However, inorganic compounds such as heavy metals are not affected by decomposition.

Operating landfills that are actively receiving garbage can be undergoing all three decompositional stages simultaneously at different locations. Normally, within the first few years following the closure of a cell, the anaerobic stage becomes dominant and remains so until all the available organic materials have been decomposed. As the landfill ages, the gas production rates gradually decrease, although the generation of landfill gas can continue for many decades. It is this long-drawn-out decomposition and methane production process that makes the extraction of methane for energy an economical undertaking.

A great deal can be learned about a landfill's contents and internal conditions by monitoring the gases and leachate. This is a potential method of studying old, closed landfills for which no records exist.

As shown in laboratory experiments, biodegradation in a landfill could be accelerated significantly by controlling and operating landfills differently than is currently done. Factors that enhance decomposition include:

- Mulching wood and lumber debris to increase their surface area
- Not compacting garbage initially, but delaying compaction for many years or decades until the decomposition is relatively complete
- Recirculating leachate through the landfill to maintain a high moisture content
- Adding chemicals and/or microorganisms to the recirculating leachate to promote biodegradation
- Aerating the landfill; this may increase the rate of decomposition but will reduce methane production

Leachate

It is ironic that the main threat to the environment from landfills that accept solid waste comes in liquid form. Some leachate is produced by organic decomposition and compaction of wet refuse, but most is generated by water infiltration from rain or snowmelt. These enter the landfill primarily during its operation before the final cap is emplaced, but they also get in in smaller amounts after closure, by leakage through the cap. This liquid spends many years percolating through the landfill. During this time it comes into contact with fingernail polish remover, paints, pesticides, motor oil, and the thousands of other items that make up the landfill. The water leaches and dissolves various constituents until it contains a load of heavy metals, chlorinated organic compounds, and other substances that can harm nearby watercourses and the underlying groundwater.

It is important to understand the quality and quantity of leachate that will be formed. Unfortunately, the processes by which leachate is generated are complex and difficult to model. Numerous factors are involved, including a detailed understanding of the water cycle, local geological conditions, waste composition, and landfill design. A number of computer models have been developed to model leachate generation. One that is in widespread use in North America is HELP, which stands for Hydrologic Evaluation of Landfill Performance (Schroeder et al., 1994). HELP estimates daily, monthly, or annual quantities of runoff, evapotranspiration, lateral drainage flux, and leakage of leachate through clay and membrane liners. Computer programs such as HELP can be used not only to help design new landfills but also to assess the environmental impact of older, closed landfills.

Although leachate composition is highly variable, it can be categorized into three distinct groups: heavy metals, dissolved solids, and organic compounds. Each of these should be addressed separately in leachate treatment systems (Sirman, 1995). The main heavy metals of concern are lead and cadmium. Sources of lead include lead-acid batteries, plastics, cans, used oil, and lightbulbs; sources of cadmium include nickel-cadmium batteries, plastics, nonfood packaging, and electronic appliances.

The chemical composition of leachate can vary greatly depending on the age of the landfill. For example, in the acid phase of decomposition, the pH value is low and the concentrations of BOD (biological oxygen demand), COD (chemical oxygen demand), nutrients, and heavy metals is high. During the methanogenic stage, however, the pH is in the range 6.5 to 7.5 and the concentrations of BOD, COD, nutrients, and heavy metals will decrease as they are less soluble at neutral pH values. In general, leachate composition changes from a higher concentration of biological materials to a higher concentration of heavy metals and total dissolved solids (TDS) as the landfill matures.

Typical ranges of the concentration of selected parameters in leachate are shown in Table 7.2. For comparison, the guidelines for Canadian Drinking Water Quality are also shown. It is seen that for almost all the listed parameters, the upper levels of the ranges of concentration exceed guidelines established for drinking water in Canada. In addition to these parameters, municipal and industrial landfills can contain other toxic and hazardous compounds, such

Table 7.2.   Typical concentration ranges of leachate from municipal landfills (mg/L) (MOEE, 1993, and Crutcher & Yardley, 1992).

Parameter	Range	CDWQG[1]
pH (no units)	3.7– 9	6.5–8.5
Alkalinity	300– 2,000	
Hardness	400– 2,000	
Total Dissolved Solids (TDS)	0– 42,300	
Chemical Oxygen Demand (COD)	150– 6,000	
Biochemical Oxygen Demand (BOD)	0– 4,000	
Total Kjeldahl Nitrogen (TKN)	1– 100	
Ammonia	5– 100	
Nitrate	<1–0.5	45
Nitrite	<1	
Sulphate (SO$_4$)	<1– 300	500
Phosphate (PO$_4$)	1– 10	
Aluminum	<0.01– 2	
Arsenic	0.01– 0.04	0.025
Barium	0.1– 2	1
Beryllium	<0.0005	
Boron	0.5– 10	5
Bromide	<1– 15	
Cadmium	<0.01	0.005
Calcium	100– 1,000	
Chloride	20– 2,500	250
Cobalt	0.1– 0.08	
Copper	<0.008– 10	1
Chromium	<0.01– 0.5	0.05
Fluoride	5– 50	1.5
Iron	0.2– 5,500	0.3
Lead	0– 5	0.01
Magnesium	16.5– 15,600	
Manganese	0.06– 1,400	0.05
Nickel	0.4– 3	
Potassium	3– 3,800	
Selenium	0.004– 0.004	0.01
Sodium	0– 7,700	200
Zinc	0– 1,350	5

[1]CDWQG = Canadian Drinking Water Quality Guidelines (Health and Welfare Canada, 1993).

as pesticides and heavy metals. It should be noted that the Canadian Drinking Water Guidelines are so strict that many natural
waters do not meet all the objectives; nevertheless, they provide a
useful point of reference.

It would seem logical that a landfill should be designed to stimulate the decomposition process so that hazardous organic compounds
will be destroyed as quickly and completely as possible. Unfortunately, this leads to a dilemma. On the one hand, decomposition

works best when the wastes are wet—that is, with plenty of leachate to cause anaerobic conditions. On the other hand, in modern landfill design, considerable effort is placed on ensuring that as little water as possible enters the landfill.

Currently there is a debate between the merits of the "wet" and "dry" landfill approaches. The wet-landfill theory holds that the landfill should be saturated with as much liquid as possible to promote bacterial growth and the biodegradation process. Leachate is collected; sometimes, treatment chemicals or microbes are added; and the leachate is recirculated into the landfill. In this way leachate is constantly circulating, and the process of biodegradation, which destroys many organic compounds, is enhanced. It is also anticipated that other inorganic compounds will be absorbed as they percolate through garbage (Pohland, 1989).

The "wet" approach is being used in the Keele Valley landfill, Canada's largest landfill, which serves the Toronto area. The intent of this "bioreactor" approach is to promote biodegradation in the early stages of the landfill, when the bottom liner has its maximum integrity. At a later stage, the landfill will be converted to dry entombment by placing an impermeable cover over it and pumping out the leachate.

In contrast, the "dry-landfill" approach has the basic premise that the drier a landfill is, the less chance it has of contaminating groundwater. An impermeable cover is an essential component of dry landfills. In addition, any leachate that forms is collected, removed, and treated. Water treatment facilities are occasionally built at such landfill sites, but more often the leachate is collected and sent to a nearby sewage treatment plant.

The process of decomposition is currently not well understood. This lack of basic knowledge, coupled with a concern about the long-term effectiveness of the bottom liner, and the enormous difficulty in remediating groundwater contamination once it has occurred, has resulted in the Environmental Protection Agency emphasizing "dry entombment." In fact, the wet type of landfill is illegal in many U.S. states.

Leachate Treatment

The most common method of managing landfill leachate involves collecting and transporting it to a sewage treatment plant. Although this approach is

→

Leachate Treatment (continued)

logistically convenient, since landfills and sewage plants are generally owned
and operated by the same municipality, it is not necessarily a sound technical solution. A study concluded that landfill leachate was not well treated by
the local sewage treatment plant (Sirman, 1995). Metals passed through the
system, and the elevated concentrations and/or toxic nature of the leachate
upset the sewage treatment process, which is based on the action of microbes
that are very sensitive to changes in the character of the influent. It was recommended that on-site treatment should be used to treat landfill leachate,
and that this should include separate processes for dealing with each of the
three main components of the leachate (heavy metals, dissolved solids, and
organic materials).

Landfill Gas: An Exploitable Resource

In addition to liquids, gases are also generated inside a landfill. Gas
management forms a key part of landfill design and operation. The
main decomposition process produces methane and carbon dioxide in roughly equal parts, as well as small amounts of benzene,
hydrogen sulfide, nitrogen, chlorinated hydrocarbons, and other
trace gases. This process lasts for many decades—even a century
or more. Typical concentrations of the main constituents of landfill gas are shown in Table 7.3.

The methane (effectively natural gas, CH_4) is flammable and is
both a danger and a blessing. Methane, when mixed in appropriate
concentrations with oxygen from the air, is explosive and can be a

Table 7.3. Typical constituents of landfill gas (Tchobanoglous
et al., 1993).

Compound	Percent (dry volume)
Methane	45–60
Carbon dioxide	40–60
Nitrogen	2–5
Oxygen	0.1–1.0
Sulfides, disulfides, mercaptans, etc.	0–1.0
Ammonia	0.1–1.0
Hydrogen	0–0.2
Carbon monoxide	0–0.2
Trace gases	0.01–0.6

serious hazard to structures on and near the landfill. On the other hand, the methane can be collected and burned in a controlled fashion to produce useful energy.

The carbon dioxide component of landfill gas is water soluble and increases the corrosiveness—that is, the acidity—of the leachate inside the landfill. This in turn leads to enhanced leaching of contaminants such as heavy metals from the garbage.

Hydrogen sulphide and mercaptans have particularly bad odors, which can be a serious nuisance to nearby residents. Many landfills have established gas collection systems, primarily to eliminate these obnoxious odors.

Improvements in detection instrumentation have allowed better understanding of the trace compounds in landfill gas, and the results indicate that many of these are harmful (Brosseau and Heitz, 1994). A total of 116 organic compounds, many of which are volatile organic compounds (VOCs), were found in landfill gas studied at 66 landfills in California (Tchobanoglous et al., 1993; see Table 7.4). The occurrence of VOCs can be related to the age of the landfill: older landfills that accepted industrial and commercial wastes contain higher concentrations, and newer landfills in which the disposal of hazardous wastes is not permitted contain much lower levels.

Many of these gases are toxic; vinyl chloride and benzene, for example, are carcinogenic. The toxicity of landfill gases has generally not been addressed in the past, and only a few jurisdictions, such as California, have introduced regulations governing the emission of landfill gases. The situation is now changing as the health hazard of these gases is becoming recognized. The U.S. EPA has determined that non-methane organic compounds emitted by landfills contribute to the depletion of atmospheric ozone, and has proposed that landfills over a certain size must install systems that collect and destroy these gases. In 1993, the EPA made it mandatory that soil gases at the perimeters of landfills be monitored quarterly.

How Do Landfill and Waste Incinerator Emissions Stack Up?

Concerns about landfills have generally been directed to potential groundwater contamination. Only recently has the issue of gaseous emissions been raised. Jones (1994) compared health and some environmental risks associated with air emissions of contaminants such as vinyl chloride, benzene, nitrogen oxides, and dioxins from landfills and incinerators of equal capacity.

→

How Do Landfill and Waste Incinerator Emissions Stack Up? (continued)

The results indicate that the health impacts are greater for emissions from landfills than from waste-to-energy incinerators, even when the landfills are equipped with gas control systems. These results are the direct opposite of both the public's perception and the focus of regulatory attention. They suggest that regulations and safety assessments should be applied to emissions from landfills similar to those for incinerators. More stringent emission standards for landfills will undoubtedly be set in the future.

Because it is in gaseous form, landfill gas is very mobile and can migrate readily, particularly if an impermeable cover prevents it from venting directly into the atmosphere. It has been known to travel through the soil as far as 1800 meters from a landfill, with the direction of travel controlled by local geological conditions. The gas can be trapped or redirected by impermeable layers such as clay and frozen ground.

Landfill gas can be collected to prevent fires or explosions caused by methane, to prevent obnoxious odors, or to extract methane as a

Table 7.4.   Typical trace constituents in landfill gas (Tchobanoglous et al., 1993); values in parts per billion, by volume.

Compound	Minimum	Mean	Maximum
Acetone	0	6,840	240,000
Benzene	932	2,060	39,000
Chlorobenzene	0	82	1,640
Chloroform	0	245	12,000
1,1–Dichloroethane	0	2,800	36,000
Dichloromethane	1,150	25,700	620,000
1,1–Dichloroethene	0	130	4,000
Diethylene chloride	0	2,840	20,000
trans-1,2–Dichloroethane	0	36	850
Ethylene dichloride	0	59	2,100
Ethyl benzene	0	7,330	87,500
Methyl ethyl ketone	0	3,090	130,000
1,1,1–Trichloroethane	0	615	14,500
Trichloroethylene	0	2,080	32,000
Toluene	8,125	34,900	280,000
1,1,2,2–Tetrachloroethane	0	246	16,000
Tetrachloroethylene	260	5,240	180,000
Vinyl chloride	1,150	3,510	32,000
Styrenes	0	1,520	87,000
Vinyl acetate	0	5,660	240,000
Xylenes	0	2,650	38,000

useful energy source. Measuring the concentration and pressure of methane and plotting the data on a map will help in determining migration patterns and planning gas extraction wells. Gas can be extracted from a landfill in a number of ways. Where off-site development is relatively close to the landfill, perimeter trenches may be dug to ensure that gases do not migrate laterally away from the site. These trenches can be designed either to act as interceptors to attract and collect gas (in which case they would contain gravel and perforated tubing), or as impermeable barriers to prevent gas from leaving the landfill site (in which case they would be filled with relatively impermeable materials such as bentonite, clay, or concrete). Alternatatively, a series of vertical wells might be installed in the landfill or along its periphery. These wells can vent directly to the atmosphere, or they can be connected together and ducted to a stack where the gas is burned (or flared) or to an energy generation station. Because of problems with odors, direct venting of landfill gas is seldom allowed.

Vinyl Chloride

Vinyl chloride (C_2H_3Cl) is a gas released in trace quantities from landfills. Colorless and with a mild, sweet odor, it is highly flammable and explosive, releasing hydrogen chloride gas on burning. It is used in the production of polyvinyl chloride (PVC) plastics and the solvent trichloroethane. There are no known natural sources.

Vinyl chloride is a serious human health risk, known to cause cancer of the liver, brain, and central nervous system. It is thought to be produced in landfills by the anaerobic decomposition of several chlorinated solvents.

In 1995, government studies at Keele Valley landfill near Toronto, Ontario, found high levels of vinyl chloride in the air, even though the landfill has an extensive system of buried pipes to collect gases for incineration (Anonymous, 1995). One half-hour air sample measured 2.9 micrograms of vinyl chloride per cubic meter, which is just under the 3.0 microgram level set for safe exposure. The gas collection pipes carried vinyl chloride at concentrations of 210 micrograms per cubic meter. The study shows that even modern landfills leak toxic gases into the atmosphere, despite sophisticated gas collection systems; this raises concerns about older and closed landfills, where gases are simply vented.

Another approach is to place a series of interconnected perforated pipes, typically of PVC, under the impermeable cover as landfill cells are closed to capture the landfill gas and direct it to collection points. Gas can be withdrawn from the landfill by applying a vacuum to the pipes. However, gas extraction should not be so excessive that air (i.e., oxygen) is drawn into the landfill. For this reason, extraction wells are generally equipped with gas sampling ports and flow control valves so that gas flow rates from each well can be properly controlled.

Landfill gas can be flared continuously, or it can be burned in a facility where thermal energy is recovered. Because of concerns over air quality, modern flaring facilities are designed to rigorous standards to ensure effective destruction of methane, VOCs, and other hazardous compounds. Figure 7.10 shows typical stacks for flaring landfill gas.

Methane production rates of 2.5 to 3.7 liters per kilogram have been reported for refuse that has been in place for a few years (Emcon, 1980). The use of methane as an energy source maximizes the extraction of useful resources from landfills and is in keeping with the principle of sustainable development. The natural gas shortage of the 1970s spurred research and development of alternate sources of energy, and one was methane gas from municipal landfills. The energy output from any single landfill is not very large,

7.10   Stacks for flaring landfill gas at Keele Valley landfill, Toronto, Ontario.

but the source will continue to generate methane for 30 to 60 years or more, and the capital investment required to construct the collection and associated systems is relatively small. Energy from the methane can be utilized in three ways:

- It can be incinerated to generate electricity.
- It can be cleaned and upgraded to a product known as "high-BTU gas" and sold to natural gas companies, which put it into their pipelines for distribution to their customers. For this option to be viable, it is necessary that a gas pipeline pass near the landfill.
- It can be sold to a nearby industry or commercial user as heating or process fuel. In this case, the gas needs to be purified, including water separation and removal of trace contaminants and particulates, but it need not be treated as much as high-BTU gas. Because of the expense of constructing pipelines, the consumers must be located close to the landfill.
- The gas can be used as a heat source for local space heating.

The first high-BTU project to become operational in North America was the Palos Verdes landfill in California, which began delivering gas in 1975 (White, 1990). In recent years it has been more economical to use methane to generate electricity than to produce high-BTU gas. In 1991, the Brock West municipal landfill near Toronto became the first landfill in Canada to generate electricity from landfill gas. At the time, its 23-megawatt electrical generating plant was the second largest of its kind in North America. At the nearby Keele Valley landfill, a 30-megawatt electrical generating plant began operation at the end of 1995.

Of the 64 landfill gas recovery projects listed in a Resource Recovery Activities report compiled in 1989, forty-five used the gas to generate electricity and eight produced high-BTU gas; among the latter was the Fresh Kills landfill in New York City (White, 1990). It is estimated that in 1998 there were more than 100 landfills in the United States that tap their methane, resulting in approximately 200 million cubic meters of fuel per year.

Because of the large number of landfills, their cumulative potential as sources of energy is relatively large. For example, if every landfill in North America were to generate electricity from methane, as much as 30,000 megawatts of energy could be produced. This would replace an enormous amount of coal, oil, natural gas, and uranium that is currently being used in electrical plants and would make a substantial contribution to conservation, reducing global warming, and preserving nonrenewable resources for future generations.

The burning of methane for energy also has other environmental benefits. Research indicates that landfills contribute between 30 and 70 million tonnes of methane per year to the atmosphere, which represents between 6% and 18% of the total methane released worldwide (Environment, 1987). Because methane is approximately 25 times more potent than carbon dioxide in trapping the sun's infrared radiation, this represents a significant contribution to long-term global warming.

To design gas extraction systems or to assess the economic viability of a methane energy system, it is necessary to know the gas quantities that will be released and how this will change over time. Barlaz and Ham (1993) calculated from theoretical considerations that one kilogram of dry waste will generate about 262 liters of methane. Their calculation assumed that cellulose and hemicellulose, the main sources of methane, constitute 51% and 12% of refuse, respectively.

Landfill gas production is a transient phenomenon that decreases with time. A number of mathematical expressions have been developed to estimate gas generation as a function of time. These generally take an exponential form:

$$C(t) = C(0)*\exp(-kt)$$

where $C(t)$ and $C(0)$ are the amounts of methane produced at times t and initially ($t = 0$), respectively, and k is a decay constant. If site-specific data are not available, then a value for k of 0.05 per year can be used (EPA, 1996). This decay is analogous to radioactive decay; for $k = 0.05$, the half-life (the time in which the amount of methane generation decreases by 50%) is about 14 years.

Environmental Monitoring

Just as a doctor must monitor a medical patient regularly for pulse rate, temperature, and blood pressure, so technicians must closely observe the vital signs of a landfill. This shows the changes that are taking place inside the landfill, providing information that will help to protect the environment. Generally, the following parameters are monitored: groundwater, surface water, landfill gas, atmospheric gas, and settlement.

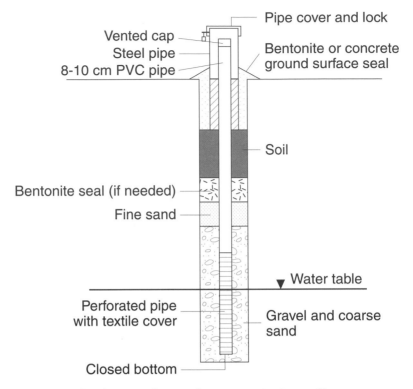

7.11 Details of a typical groundwater monitoring well.

Water Monitoring

Groundwater depth, flow direction, quantity, and quality can be determined by installing a network of monitoring wells (see Figure 7.11) and sampling them at regular intervals. A typical groundwater monitoring network is shown in Figure 7.12. A modern landfill may have from 20 to more than 100 monitoring wells around it, depending on the complexity of its geology. A key part of the monitoring program is to establish background, or baseline, water quality information, so that any perturbations caused by landfill leachate can be recognized. The direction of the groundwater flow should be determined and any nearby homes, farms, and water wells that might be affected by leachate should be identified. Because the chloride ion is common in leachate (but not in groundwater) and because it can be detected relatively easily, it has proven to be an excellent tracer of leachate in groundwater.

Surface waters, such as nearby streams or ponds, are monitored periodically by direct sampling.

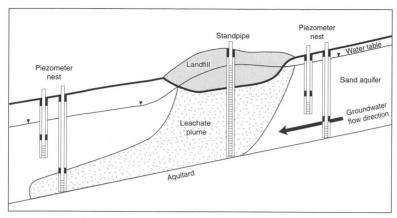

7.12   Typical monitoring well network, shown in cross-section.

Landfill Gas Monitoring

This keeps track of the movement and location of landfill gas. Probes are placed below the landfill cover to measure gas pressure, temperature, and composition. In addition, probes are placed around the site to detect any gas escaping from the site.

Atmospheric Gas Monitoring

Most modern landfills monitor the quality of air at a number of points around the perimeter of the property. This is of greater importance at landfills that do not collect and burn landfill gas. Most gas sampling devices are either grab or active samplers. Grab samples are collected using a collection bag made of an appropriate synthetic material, an evacuated flask, or a gas syringe. An active sampler collects and analyzes a continuous stream of air. As with groundwater and surface water, it is important to determine the background, or ambient, air quality as a basis for comparison.

Settlement

It has become standard practice to run a survey-control benchmark elevation circuit on and around landfills to provide references for measuring settlement. Plates set in fill with pipes extending to the surface, or concrete or stone monuments 15 centimeters square and one meter long, make good survey markers. Standard engineering surveying instruments are used, and surveys are run periodically, usually at monthly or quarterly intervals.

Closure and Post-Closure Care

To ensure that landfills will be maintained after they have been closed, most jurisdictions have passed regulations requiring owners of landfills to prepare a detailed post-closure plan and to establish a post-closure fund. Most post-closure programs are a few decades in duration.

The most important elements of a post-closure plan include:

- Cover and landscape design.
- Control of landfill gases. This may include continued gas collection after closure.
- Environmental monitoring systems including groundwater and air quality.
- Collection and treatment of leachate.
- Ongoing inspections and maintenance.

Because most landfills protrude well above the surrounding landscape, they are exposed to erosion by wind, rain, and snow. These forces, aided by freeze/thaw cycles, animal burrowing, and landfill settlement, will eventually—in decades or centuries—compromise the integrity of the cover system and allow water to enter. Regulatory agencies and other experts agree that even the best liner and leachate systems will ultimately fail as a result of natural deterioration (Lee & Jones, 1991). At best, such systems only postpone groundwater pollution. To avoid this problem, landfills will require monitoring and ongoing maintenance for centuries to come. This places a responsibility and burden on future generations. Because landfill heights have grown enormously in recent years, it is often impractical to excavate through tens of meters of landfill to make repairs, even when leaks are discovered.

Innovative Technologies in Remediating Old Landfills

A major environmental problem is presented by the tens of thousands of old, closed landfills that were constructed without liners and leachate collections systems. One innovative approach to this problem was demonstrated at Livingstone County landfill near Pontiac, Illinois (Yach, 1996).

The landfill began operation in 1978 and now occupies 20 hectares (50 acres), reaching a height of 25 meters, the height of an eight-story building. A segment of about 13 hectares (32 acres) of the site was constructed with dual

→

Innovative Technologies in Remediating Old Landfills (continued)

composite liners and a leachate collection system, which was designed to maintain the depth of leachate above the liner at 30 centimeters or less. Although the older part of the landfill (about 7 hectares, 18 acres) has a liner, it did not have a leachate collection system. The depth of the leachate in this section of the landfill ranged from 1 to 12 meters.

Instead of using a grid of vertical wells, engineers decided to install a single 210-meter horizontal well that would be able to drain a much larger portion of the landfill and would require less pumping energy—an approach that had never been attempted before in North America. The bore would start at the east end of the old site and angle upward at the slight grade of 0.35%, so that leachate would flow back to the east side, where it would be collected and piped to a storage tank, with periodic removal to the local sewage treatment plant.

The project presented several technical difficulties. First, the landfill was very heterogeneous and full of many different kinds of waste including concrete, steel, and unknown materials. Second, the horizontal hole was to be drilled directly above and as close as possible to the liner so that the maximum amount of leachate would be collected; accurate navigational control was vital since the drill head must not deviate downward and puncture the liner. A three-dimensional sensor package was used to navigate the drill head, transmitting position and orientation data every three seconds down the drill piping. The system worked well, and the drill hole emerged only 30 centimeters higher than planned. A 20-cm hole was bored and then a 10-cm perforated HDPE pipe was pulled into place.

Instead of the usual bentonite clay, which would plug the perforations in the leachate collection pipe, a special drilling fluid was used that had sufficient gel strength to carry out drill cuttings and support the hole, yet that would decompose over time so as not to impair leachate collection.

Human intrusion is a significant future risk, because humans are by nature curious. It is inconceivable that a large mound will not appeal to the exploratory instincts of future generations. Furthermore, landfills are located in close proximity to urban centers, and as communities grow and expand there will be an increasing pressure to build on closed landfills. Thus, landfills will require long-term guardianship. The prevention of future intrusion into landfills is currently not addressed in the regulations or in landfill design.

Landfilling of garbage is underpriced. Landfill cost estimages do not include the future despoliation of the environment or the care and maintenance that must be performed by future generations.

Future Developments

What will influence the future evolution of landfills? First and foremost are the ongoing programs by all levels of government to divert waste from landfills through waste reduction at source and recycling. Increased emphasis is being placed on blue-box programs and composting of yard and other organic materials. These programs are critically important in finding a solution to the garbage crisis.

As the volume of waste requiring disposal diminishes, this process should have a major impact on how landfills are designed and operated. Since less waste will be disposed, it should be possible to design higher-quality and more durable landfills, even though they are more costly.

In addition, the organic content of waste may decrease because of the diversion of paper and plastics by blue-box programs, composting of yard wastes, and, in particular, incineration. A reduction in organic content would influence the decomposition processes that occur and could lead to smaller methane yields as well as slower rates of decomposition. For example, the waste would tend to be more compact and denser, affecting the ways liners, gas collection systems, and other engineering components of the landfill are designed.

Another major trend is the evolving understanding and control of the biological decomposition processes inside landfills. Experiments are being done in collecting leachate and recirculating it through the waste in a controlled manner. The intention is to operate landfills as controlled bioreactors, much as modern sewage treatment plants actively promote biological and chemical action to break down liquid wastes. Chemicals may be added to the leachate recirculation system to promote more complete degradation.

This approach also has promise for managing inorganic compounds. For example, acidic compounds naturally produced in landfills leach out heavy metals. By controlling the process, sulfate compounds can be reduced to sulfides which precipitate heavy metals. If leachate is appropriately treated and recirculated, heavy metal precipitates could be bound up in the bulk of inactive landfill material. Alternatively, they could be removed by appropriate treatment techniques before the leachate is returned into the landfill. Although such methods show promise, they are still in the developmental stage.

There is also steady improvement in the quality of the materials used in landfill construction, such as better liners and better methods for sealing their seams.

In summary, the disposal of our society's wastes has been and continues to be tied inextricably to the basic landfill. The thousands of landfills that dot the country symbolize our profligate, wasteful, throw-away, over-packaged, and over-marketed consumer lifestyle. Although many engineered bells and whistles are now common-place—liners, impermeable covers, and leachate and gas collection systems—the basic concept of disposing of waste in near-surface dumps close to urban centers has remained largely unchanged for centuries. With population growth and increasing environmental concern, landfills will continue to evolve. In the coming years there will be a trend toward decreasing the amount of waste going to land-fill, which in turn will lead to a change in the composition of the waste emplaced. Further improvements are expected in the design of engineered barriers and systems.

Discussion Topics and Assignments

1. Who do you consider most qualified for assigning weighting and suitability ratings for landfill site selection criteria? Should people who are not professionals be involved in the process? Why?

2. The following criteria have been developed for the siting process for a landfill in your area. What weights would you give each one (5 is maximum)? Give your reasons. (a) existing land use and zoning, (b) distance to access roads, (c) noise impact on nearby residential areas, (d) depth to groundwater table, (e) visual impact, (f) distance to waste source area.

3. Calculate the area needed per year and for a 30-year lifetime for a new landfill for a region with 400,000 inhabitants that generates 2 kg/day/person of solid waste. Don't forget the buffer, storage, and administration areas, which will require 25% to 35% of the total area.

4. Calculate the amount of daily cover soil required for the following hypothetical landfill. Also, calculate the waste-to-soil ratio. 70 tonnes of waste are received each day; each lift is 3 m high; cell width is 5 m; the slope of the working face is 1:3; compaction is to 350, 475, and 600 kg/m³; daily

cover is 15 cm. Hint: You need to consider a 3-dimensionsal trapezoidal volume—i.e., surface area to receive soil cover is top, front, and side.

5. What is the breakthrough time for leachate to penetrate a 1.3 meter clay layer (d) if porosity (por) is 0.2, coefficient of permeability (K) is 10^{-7} cm/s, and hydraulic head (h) is 2 m? Use the formula:

$$Time = (d^{2*por})/K*(d+h).$$

6. Locate a closed landfill in your area. What is being done to monitor it, and what measures have been instituted to ensure that gases and leachate do not harm the nearby environment? How long must these measures stay in place?

7. A landfill receives municipal waste at the rate of 2,000 tonnes per day for 20 years. Using the data presented in this chapter, what is the theoretical maximum amount of methane that will be generated? What is the initial gas generation rate, once the methanogenic stage begins?

Suggested Reading

Henry, J. G. 1989. Solid Wastes. In *Environmental Science and Engineering*, J. G. Henry and G. W. Heinke (eds.). Englewood Cliffs, N.J.: Prentice-Hall.

McBean, E. A., F. A. Rovers, and G. J. Farquhar. 1995. *Solid Waste Landfill Engineering and Design*. Englewood Cliffs, N.J.: Prentice-Hall.

Ministry of Environment and Energy. 1993. *Guidance Manual for Landfill Sites Receiving Municipal Wastes*. PIBS 2741. Ottawa.

Murphy, Pamela. 1993. *The Garbage Primer: A Handbook for Citizens*. New York: League of Women Voters Education Fund, Lyons & Burford.

Rathje, W. L., and L. Psihoyos. 1991. Once and Future Landfills. *National Geographic*, May, pp. 116–134.

Rathje, W. L., and C. Murphy. 1992. *Rubbish! The Archaeology of Garbage*. New York: Harper Collins.

ARE THERE BETTER
DISPOSAL METHODS?

Most of the solid waste generated by society ultimately winds up in near-surface landfills. Let us put our thinking caps firmly on, place our prejudices aside, and explore what other methods might be used to dispose of waste. We should seek, in particular, the approaches that best fulfill the three basic principles described in chapter 2. That is, we should strive to find disposal methods that are in accord with sustainable development.

Existing and Abandoned Mines: The Hole Is Already There

Existing and abandoned pits, quarries, and mines are attractive for waste disposal because a hole to contain the wastes has already been excavated. Such abandoned areas, when left unreclaimed, cannot be used for agriculture or other beneficial uses. Thus, they generally do not have significant market value and can often be obtained relatively cheaply. For these reasons, pits and quarries have been extensively used for landfills. Operating and abandoned mines, on which this section focuses, are somewhat similar to pits and quarries, though usually larger.

Abandoned mines hold promise as disposal facilities because they are resource areas that have been depleted and thus have little future value. There are two basic types of mine: the open pit mine, which is effectively a large pit or hole in the ground; and the underground mine, where the mined-out openings are deep underground and there is no surface expression except for the shafts used to gain subsurface access. Because underground mines occupy minimal surface land, their use for waste disposal would be in accordance with the sustainable development principles that were advocated in chapter 2. Several European countries, with higher population densities and much smaller land mass than in North America, have long used abandoned underground mines to dispose of their rubbish.

The major advantage of placing wastes deep in underground mines is that it is inherently safer than placing the wastes in a surface facility. The amount of groundwater and its flow rate decrease with depth; this fact, combined with the long transport paths back to the biosphere, minimizes the possibility that contaminants will be carried by groundwater to the surface, where they could damage the environment. The waste is contained deeper and more securely.

Furthermore, an underground waste facility is removed from the forces of erosion and does not require ongoing maintenance. It is also out of sight and not as attractive or obvious a target for intrusion as a surface landfill. For these reasons, an underground mine will not place as great a burden on future generations as a surface landfill.

Abandoned open pit mines also have some advantageous features, and a number of proposals have recently been put forth for their use as municipal disposal sites. One such case history is described in chapter 11, with an aerial view of an abandoned open pit mine shown in Figure 11.8. Open pit mines are particularly despoiled pieces of land that not only are scars on the landscape but also pose safety hazards. They can pose environmental threats as well: for example, sulphur-containing ores and tailings can generate acid water in the bottom of the abandoned pit, which can contaminate nearby water bodies. Converting an abandoned open pit mine into a landfill can alleviate some of these problems, namely, the area will be more attractive visually and less acid water will be generated because the landfill will have water-tight bottom liners.

Nonetheless, there are also potential drawbacks to the use of mines for waste disposal. It must be borne in mind that the primary purpose of a mine is resource recovery, not waste disposal. This means that ore extraction is maximized while the structural stability of the mine is kept at a minimum. In many cases, underground support structures such as pillars are removed during the final stages of operation. As a result, the underground openings and shafts become unstable and may collapse. The same is true of open pit mines, where slopes are designed as steep as possible to maximize ore production. In other words, abandoned mines have been designed to stay open just long enough to extract their mineral content—not for an additional lifespan of 20 or 30 years for waste disposal.

Another potential shortcoming is that mines are often situated in areas of faulting and/or folding of the rocks which in the geologic past have formed conduits and traps for the mineralization that is being exploited. Thus, mine areas are often geologically complex and contain many faults, which can be excellent conduits for groundwater flow. In such instances, an abandoned mine might need additional engineered barriers to control contaminant migration. The complexity of the geologic setting would also make it difficult to develop the predictive computer models of groundwater flow and contaminant transport that may be required for licensing.

A further potential detriment of using abandoned mines is the possibility that the remaining low-grade ore (and possibly the tailings) may at some future time become economical to mine because of price changes and/or new technologies. In such a case, the use of the site for waste disposal would prevent or cause difficulties to future miners in pursuit of this resource.

For these reasons, not all abandoned mines are suitable for waste disposal. A detailed, site-specific technical assessment of each mine is necessary.

Although very few abandoned mines have been used for municipal waste disposal in North America, both underground and open pit mines have been used for disposal of another waste—mine tailings, the residual rock remaining after the mining and milling process. At the Elliott Lake uranium mines in Ontario, tailings have been placed underground in a cement-slurry matrix. The slurry allows the large volumes of material to be transported using pipelines. This is an important consideration for underground facilities because the wastes must be transported through narrow shafts and tunnels. Underground disposal is more environmentally responsible than placing the tailings in surface piles. In addition, the tailings, when mixed with cement, help to stabilize and support the mined openings, allowing ore to be extracted from the remaining pillars and support columns, which would otherwise not have been mineable.

In Germany, mines have been used for disposal of radioactive wastes. Between 1967 and 1976 about 62,000 containers of low-level radioactive waste were deposited in the Asse mine, an abandoned salt mine that had been worked for almost a hundred years (Tammemagi & Thompson, 1990). Salt mines have the unique characteristic that the salt is a plastic material; that is, it slowly creeps over time so that the mined openings slowly squeeze shut, creating a seal around any emplaced wastes. The very existence of such salt deposits, which formed hundreds of millions of years ago, attests to the dry nature of the surrounding sedimentary layers, because salt is water-soluble and would long ago have dissolved if any groundwater had been present.

Hazardous Wastes Go Deep Down

Since 1972, an old salt mine near Heringen, Germany, has been used by Kali und Salz AG to store more than 2 million drums of solid hazardous waste. The disposal is at a depth of 700 meters in salt formations that have been stable for 250 million years and, presumably, will remain so for eons to come. Deep, dry, and geologically stable, the facility is considered to be one of the world's safest (Boraiko, 1985).

The disused Konrad iron ore mine in Lower Saxony, Germany, with workings between 800 and 1300 meters in depth, is currently being used for radioactive waste disposal (Tammemagi & Thompson, 1990). The mine, which is very dry owing to about 800 to 1,000 meters of overlying, predominantly clay rocks, has the capacity to receive approximately one million cubic meters of low-level radioactive waste over its 40-year lifespan. Figure 8.1 shows a schematic cross-section of the Konrad mine.

Although this discussion has emphasized using *abandoned* mines, other options are also available. In underground mines, it may be possible to combine ore extraction operations with simultaneous waste emplacement. The latter would take place in mined-out sections of the mine and may require a dedicated shaft. This combined mode of operation has certain advantages for waste emplacement in that the infrastructure is already in place. If the waste disposal facility includes an incinerator, it may be possible to dispose of incinerator ash in a cement slurry, much as tailings have been placed underground. This approach may be beneficial to the mining operation by helping to stabilize underground openings that might otherwise collapse.

Another option would be to construct an underground facility specifically for solid waste. Although this has not been done to date in North America because of economic considerations, it may become feasible in the future as waste quantities decrease. This

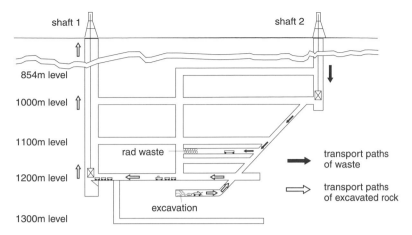

8.1   Cross-section of the Konrad Mine, Germany.

approach has the advantage that the waste disposal mine could be sited in a location and geologic setting that maximize safety. This concept has been applied to low-level radioactive wastes in Sweden; a case history is described in chapter 11. Sweden also uses subsurface disposal for hazardous wastes, first encapsulating them in concrete and then placing them in underground vaults (Miller, 1997).

Open pit mines have also been used for disposal of mine tailings. This is similar to the use of pits and quarries for municipal landfills. An innovative approach has been developed at the Rabbit Lake uranium mine in northern Saskatchewan in Canada (Clark, 1989b). Their in-pit disposal system was designed to take advantage of the fact that the pit walls are relatively permeable to groundwater, (as a result of blasting), compared to the underlying ground. A system was designed in which the groundwater is induced to flow through the permeable walls and engineered envelope rather than through the compacted and relatively impervious consolidated tailings. This disposal system is known as the "pervious-surround" or "flow-around" system.

The Rabbit Lake system promoted the consolidation of tailings during placement by draining water from the tailings. The water was collected by a tunnel from the bottom of the pit which leads to a sump pump station and a shaft to the surface. The permeability of the pit walls was enhanced by lining the walls with a permeable sand and rock layer. Figure 8.2 shows a schematic cross-section of the flow-around disposal system. When the placement of tailings

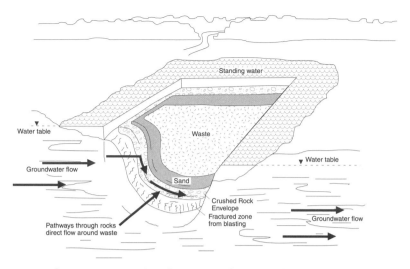

8.2   Schematic view of a "flow-around" waste disposal system.

is complete and the monitoring of the discharge from the pump station shows satisfactory results, the pumps will be removed, and the contrasting permeabilities of the tailings and the surrounding wall and envelope will establish a groundwater flow pattern that will entirely bypass the tailings.

Landfill Mining: "Play It Again, Sam"

Landfill mining, also known as landfill reclamation, is a recently developed method for extending the lifespan of old dumps. Landfill mining is simply the excavation of an existing landfill and the removal from it of various materials for the purpose of recovering, reusing, or recycling them. Items that can be usefully removed include appliances (known as "white goods"), compostable materials, and soil cover. Landfill mining has the greatest impact on reducing waste if it is operated in conjunction with a waste incinerator; that is, the excavated garbage is used as fuel for a waste-to-energy incinerator.

The materials that are to be replaced in the landfill can be shredded and compacted to minimize their porosity and decrease the total space required. As well as freeing up valuable landfill space, the reclamation process allows environmental upgrades to be implemented.

Landfill mining was pioneered in 1988 in Collier County, Florida. The objectives of the project were to reduce the potential for ground-water contamination, recover and reuse cover material, and reclaim landfill capacity. As a result of the success of the project, Collier County has included landfill mining as an integral part of its waste management strategy.

Landfill mining is also being successfully practiced in southern Pennsylvania at the Frey Farm Landfill, which services Lancaster County, consisting of 60 communities with a population of 422,000 (Flosdorf, 1993). The landfill was opened in 1989 and occupies 62 hectares (153 acres), of which 39 hectares (96 acres) is occupied by seven disposal cells and the rest is buffer area. The total capacity of the landfill is 7 million cubic meters. The landfill has a double liner, of which the bottom, or secondary, liner is made of a composite system of clay and a 1.5 mm (60-mil) high-density polyethylene (HDPE) liner. The top or primary liner is also a composite section, with 1.5 mm (60-mil) HDPE and bentonite subliner. Above the primary and secondary liner systems are highly permeable leachate collection zones. The primary system empties into a pump station which transports the leachate to an on-site leachate treatment plant, while the secondary system empties into a manhole used to detect and monitor any leakage in the system. It was originally estimated that the landfill would be filled to capacity in 9 years, even with an aggressive recycling program. The landfill's closure plan calls for the development of a natural and wildlife reserve area including a hiking trail, which is compatible with the surrounding farmland and woodland.

An important goal of the Lancaster County Solid Waste Management Authority was to protect farmland, one of the county's major resources, by extending the life of the existing landfill as much as possible. Therefore, an electricity-generating incinerator with a capacity to receive 1,100 tonnes of waste per day was constructed and began operation in 1991. It produces approximately 35 megawatts of electricity. This incinerator is described in chapter 11.

Landfill mining began about a year later with the following objectives:

- To reclaim landfill space, thus extending the life of the landfill.
- To increase energy production and efficiency at the waste-to-energy facility by increasing combustion capacity.
- To recover valuable resources such as metal and landfill cover material.

The system evolved over the initial years. A significant improvement was the introduction of a trommel to screen out soil from the waste. The trommel consists of a 2.1-meter diameter rotating drum 7.0 meters long. Excavated waste is fed into the trommel, and the 2.5-cm openings effectively separate soil from the refuse. A view of the trommel in operation is seen in Figure 8.3. About one tonne of cover material is recovered for every four tonnes of waste excavated. This results in a cleaner fuel with higher heating value, less soil in the incineration ash, and more recovery of reusable landfill cover material. The cover material is a useful resource because state regulations require that an active landfill face must be covered with at least 15 centimeters of soil at the end of each working day.

Fresh waste is mixed with reclaimed waste in a ratio of about 3 to 1. About 4% to 5% by weight of chipped tires and shredded wood is added to the waste to improve the heating value. On average, about 2,200 tonnes of material are excavated from the landfill each week. Of this, 68% is used as fuel for the waste-to-energy plant; 28% is used as cover material; and only 4% is noncombustible and is returned to the landfill. Of the material sent to the incinerator, approximately 32% becomes ash, which is returned to the landfill. Thus, 75% by weight of the waste removed from the landfill is reclaimed by the mining process.

8.3   Landfill mining at the Frey Farm landfill, Lancaster County, Pennsylvania (courtesy of Lancaster County Solid Waste Management Authority).

This is a very significant saving in landfill capacity and will extend the lifetime of the facility fourfold. Without incineration or mining, the landfill would have become full in 9 years, and a new landfill would have been required by 1998. Now the landfill can operate well into the next century providing an extra 20 to 30 years of service. Over this period, Lancaster County will avoid building an additional two or three landfills, thus saving about 120 to 160 hectares (300 to 400 acres) of its prime farmland.

Economic benefits include electricity sales from the waste-to-energy incinerator, the sale of recovered iron and other recyclable materials, the recovered material for daily landfill cover, and—probably the most significant item—the value of the reclaimed land-fill volume. Cost analyses show that the operation is financially successful.

With this cutting-edge approach to waste management combined with a proactive recycling program, only 15% of the 365,000 tonnes of garbage produced in the county each year winds up in the land-fill. This is markedly better than the U.S. national average of about 60% and the Canadian average of about 70%. (The percentage of municipal garbage going to landfill in the United States is less than in Canada because incineration is used more widely; the contribution of recycling is about the same in both countries.)

Along with freeing valuable landfill space, environmental up-grades can be implemented during the reclamation process, such as removing hazardous materials, installing synthetic or clay liners, and installing leachate and/or methane collection systems. For example, at the Trail Road Landfill, which serves the city of Ottawa, Ontario, managers are considering mining older parts of the landfill where liner systems were not originally installed. In addition to extending the lifetime of the landfill, the mining operation would offer the opportunity to construct a proper liner and leachate collection system, resulting in a landfill that is safer for the environment.

With the strong opposition to siting new landfills, landfill mining may become a valuable tool for extending the lifespans of existing landfills. The practice is now being successfully used at a small but growing number of sites throughout North America.

But landfill mining holds even greater promise. Vigorous research is being conducted to find safe and beneficial uses for incinerator ash, such as road aggregate, construction blocks, underwater reefs and seawalls, and more (Woods, 1991). Should this research lead to viable, commercial, and safe markets for incinerator ash, then the garbage that is being sent to the Frey Farm landfill could be reduced to less than 5% of the total produced in the county. Another landfill would not be needed in Lancaster County for more than 150 years!

We are beginning to see a glimmer of hope that there are solutions to monster landfills and their attendant environmental problems. The citizens and environment of Lancaster County would not be the only ones to benefit from the advances pioneered there.

Ocean Dumping: "Out of Sight, Out of Mind"

Oceans today are not the infinite, self-cleansing resource they once were. On the contrary, the rapid growth of the world's population to 6 billion people has exposed just how finite the oceans are. Oceans are the ultimate sink for much of the waste humans produce. The world's oceans receive agricultural and urban runoff from rivers, atmospheric fallout, garbage and untreated sewage from ships, and accidental (and occasionally intentional) oil spills from tankers and drilling platforms. In addition, industrial wastes, sludge from sewage treatment plants, and materials dredged from the bottoms of rivers and harbors are often dumped in the oceans. There are many signs that the oceans are under considerable stress.

The Baltic Sea, for example, is surrounded by 70 million people and their industrial effluent. The sea has been overloaded with pesticides, toxic industrial wastes, and excessive nutrients in the form of nitrogen and phosphorus from municipal sewage and agricultural fertilizers. It is estimated that the Baltic receives approximately 2.3 million cubic meters per day of domestic sewage, of which 40% is untreated (Clark, 1989a). Many of its beaches are closed to swimming, and its deeper layers have become so depleted of oxygen that some areas are devoid of life. Eutrophication—algae blooms and other plant growth that disturb the normal gas balance—has become a significant problem owing to the large amounts of nitrogen and phosphorus runoff from industry and agriculture. In the 1960s, high concentrations of mercury occurred in fish along coastal areas; commercial fishing was banned in Sweden where the fish contained more than 1 ppm of mercury. The cause was found to be mercurial pesticides, primarily from pulp and paper companies. Once the source was discovered, actions were taken to reduce the use of such pesticides. Despite a ban on the use of PCBs, there are still land-based inputs into the Baltic, and the contamination of fish and guillemot (a seabird) eggs has not shown reduction since

1974 (Clark, 1989a). PCB concentrations in fish are unduly high; the main threat is to fish-eating birds and mammals such as birds of prey and seals.

Many other seas, including the Gulf of Mexico, the Mediterranean Sea, the Arctic Ocean, Chesapeake Bay, and the North Sea, are also seriously imperiled by pollution.

A symbol of how our garbage is despoiling nature is provided by the plastic bottles and other debris that have been reported on the beaches of uninhabited atolls in the South Pacific and floating far out to sea. Small pellets (3–4 mm in diameter) of polyethylene, polypropylene, and sometimes polystyrene are widespread in the oceans. They have been reported since the early 1960s, mostly near shipping lanes or centers of industry but also in the South Atlantic and South Pacific, thousands of kilometers away from any industrial source. Larger plastic debris occurs in abundance in the oceans, and it is estimated that much of it arises from ships throwing garbage overboard (Clark, 1989a). Plastics are not biodegradable and have a long life in the sea.

Oceans are a natural target for waste disposal primarily because of their seemingly endless expanse. People imagine that there is an almost limitless dilution available which should negate the harmful effects of any contaminants. The remoteness of the oceans is also a contributing factor: no human bystanders can witness dumping, and any wastes that are dumped are "out of sight and out of mind." In addition, the open seas and oceans are international territory, and although international treaties provide some regulatory protection, there are few effective means of enforcing these guidelines. Thus, there is a natural tendency for the unscrupulous to bend the rules.

> "The very survival of the human species depends upon the maintenance of an ocean clean and alive, spreading all around the world. The ocean is our planet's life belt."—— The late marine biologist and explorer Jacques Cousteau

Before the turn of the twentieth century, much of New York's solid waste was simply dumped at sea. This proved a cheap and convenient solution until beachfront communities began to protest against the garbage washing ashore. This practice was discontinued in about 1900 (Walsh, 1991).

Dumping garbage from ships and barges, however, continues to be a standard practice of many countries throughout the world. In addition, oil spills and petroleum contamination from oil tankers, seabed oil-drilling projects, and the large amount of marine traffic has polluted coastal areas and large tracts of the ocean. The Soviet Union used the Arctic Ocean for dumping of radioactive wastes, including nuclear reactors from their nuclear-powered fleet. Immeasurable damage has been done to marine life.

The practice of dumping the sludge from sewage treatment plants at sea has continued for many decades. In 1977, the U.S. Ocean Dumping Act was amended to ban the dumping of sewage sludge by the year 1981. New York City, having no alternative land sites for its sludge, fought the act and was able to continue for several years dumping approximately 6 million tons of sludge into an area 185 kilometers offshore. Adverse effects on marine life prompted the U.S. Congress to pass the Ocean Dumping Act of 1988, which became effective in 1991 and prohibited the ocean dumping of any wastes from U.S. sources.

An innovative scheme to avoid the difficulties of the NIMBY syndrome has been the use of ships as floating incinerators (Boraiko, 1985). Several European nations burn wastes in incinerators on ships, with the subsequent ash being dumped at sea. Exhaust gases are not scrubbed; hydrochloric acid, the main product, is rapidly absorbed by the sea, which has an enormous buffering capacity. By avoiding expensive scrubber systems, incinerator ships can dispose of waste significantly more cheaply than land-based incinerators. During the 1980s, about 100,000 tonnes per year of chlorinated waste, two-thirds of it from Germany and the rest from other European countries, were incinerated at a designated site 70 nautical miles off the Dutch coast in the North Sea. The combustion temperature exceeded 1200°C and achieved 99.9999% destruction of PCBs. The practice was regulated and closely monitored by the Dutch authorities. Despite the advantages of ocean incineration and the lack of evidence indicating detrimental effects on the marine environment, the practice was phased out in 1995 (Clark, 1989a).

In the United States, Waste Management, Inc., one of the giants of commercial waste management, was allowed to conduct experimental burns of liquid hazardous wastes in the Gulf of Mexico. In 1991, as a result of protests against the practice, the EPA ordered an end to the incineration in the gulf.

In 1972, 90 nations met in London and negotiated the London Dumping Convention, which regulated pollution from ships, oil rigs, and other marine sources. The dumping of certain toxic substances—such as cadmium and mercury compounds, as well as high-level nuclear wastes—was banned. Until 1971, 8 European countries used 10 sites in the northeastern Atlantic for dumping low-level radioactive wastes such as contaminated piping, concrete and building material, glassware, and protective clothing, from nuclear power stations, universities, and research centers. The waste was packed in concrete-lined steel drums and embedded in bitumen or resin to ensure that the containers did not implode from the great pressure at the seabed. The steel, concrete, and bitumin/resin delayed corrosion and leaching of the contents, allowing the radioactivity to decay and ensuring a slow release of the contents so they would be greatly diluted (Clark, 1989a).

After 1971, the dumping sites were combined into one site 900 kilometers offshore and 550 kilometers beyond the edge of the Continental Shelf, with a depth of 4400 meters. The United States used a site 2800 meters deep in the western Atlantic between 1946 and 1967. In 1983, the London Convention passed a resolution that called for an end to the dumping of low-level nuclear wastes, and this practice ceased.

The International Convention for the Prevention of Pollution from Ships was expanded in 1978 to ban the dumping of various types of waste, including garbage bags, fishing nets, and ropes. The U.S. law that implements the International Convention includes provisions that forbid dumping of plastics within 320 kilometers of the coast and the dumping of other wastes such as metal and glass within 19.2 kilometers. In 1991, another provision was added to the International Convention that declared the Gulf of Mexico a Special Area and banned dumping of any kind there.

Uncontrolled dumping at sea is clearly not acceptable. The view that no wastes at all can be disposed of at sea is becoming more prevalent. If sea disposal is ruled out, however, those wastes will have to be disposed of on land, also an unattractive option. The natural question is whether there is some happy medium. Are there some classes of waste that could be dumped in the ocean in controlled quantities without causing detrimental effects?

The oceans have considerable capacity to absorb organic wastes, which are subject to bacterial attack that decomposes these compounds to stable inorganic compounds such as carbon dioxide, water, and ammonia. In principle, such degradable wastes are no different from the plant and animal remains and excreta that occur naturally in the marine environment. The problem then is twofold:

first, to ensure that any organic wastes that are disposed do not contain other contaminants—such as heavy metals or pathogens—in quantities that are deleterious; and second, to ensure that too much waste is not disposed at any one site, so that the local area does not become overloaded, resulting in oxygen depletion.

Experiments have been conducted with dumping compacted bales of shredded refuse to determine whether they would act as suitable substrates for marine life (Loder et al., 1983; described in Neal and Schubel, 1987). Some of the bales contained food waste and others did not. Glass and metal were added to increase the density and ensure that the bales would sink. Mobile organisms increased in abundance on and near the bales, but no changes were observed in the animals that lived on the bottom. The bales maintained their integrity for the year they were studied, and concentrations of potentially toxic products formed within the bales dropped to ambient levels in the nearby seawater and were not deemed to pose a threat to marine life.

Experiments have also been conducted that incorporated incinerator ash in concrete blocks used to build artificial marine reefs that would provide a habitat for fish. An artificial reef was constructed by researchers at the State University of New York at Stony Brook in Conscience Bay off the north shore of Long Island, using large blocks (3 m by 1.2 m by 0.6 m) containing 65% to 75% ash from three waste incinerators (Woods, 1991). After five years, the reefs were still very stable and showed no signs of leaching.

Carefully controlled disposal at sea of limited types of waste, in limited quantities, in a manner that is sustainable, should be achievable. But defining the standards requires considerable scientific research and understanding of the marine environment and the impact of the wastes on it.

The issue of ocean dumping of wastes has a sense of "déja vu" about it. At first, humans thought that land was limitless and that garbage disposal could be done with impunity. If necessary, we could just move to a new locale; space was not a problem, and the adage, "Out of sight, out of mind," ruled the day. We have learned that this is not true: with a fast-growing population, our land and associated surface and groundwaters have become incredibly precious resources. Yet we are now repeating the same mistake with the oceans. Unlimited and careless ocean dumping violates the basic principles of sustainable development. The oceans are not an infinite and limitless resource. Unless we cease careless disposal, we will eventually despoil the oceans, our last frontier.

Deep-Well Injection

The disposal of liquid wastes by injection into wells that penetrate deep into subsurface geologic formations has become relatively common in some areas. This practice was originally developed for disposal of brines produced as byproducts of oil and gas exploration and production. The method has been used for particularly hazardous wastes, such as acid and cyanide compounds, pesticide residues, process wastes, and even military nerve-gas byproducts. The justification is that the wastes are injected into carefully selected geologic formations that are so deep (the wells can penetrate two kilometers or more), so hot (temperatures at these depths can reach over 200°C), and so laden with natural dissolved chemicals that their groundwaters would be of no conceivable use to humans for drinking, irrigation, or industrial uses.

Deep-well injection is still in use for hazardous wastes today (see, for example, chapter 11, the Alberta Special Waste Management Facility). This technology is feasible in locations that have deep sedimentary geologic formations and also where oil exploration and pumping is going on, so that the drilling equipment and necessary well-development expertise are available. Most injection disposal wells in the United States are in the Texas and Louisiana petroleum fields, whose geology has been very well mapped.

Underground geology is critical to the siting of a deep-injection well. The target injection zone(s) must be isolated or confined below stable formations and must be removed from areas of petroleum or gas production, as well as from any aquifers that could be used for water extraction. In fact, suitable injection zones should lie below any areas of petroleum production in a formation that has no conceivable use—for example, one that contains saline or brackish water—and that has no connection with shallower formations containing aquifers. The target formation also has to have appropriate hydrologic properties to be able to receive the intended waste volumes.

Deep-well injection is a high-technology disposal method employing equipment and methods from the petroleum and gas industry. The injection well must be designed to withstand the corrosive conditions and high temperatures that may be encountered at depth. The well will almost certainly pass through aquifers or petroleum reservoirs, and so it must be constructed with one or more casings, leak-detection systems, and formation-sealing systems to prevent leakage (EPA, 1985).

In spite of precautions, accidents can occur. In 1984 in Ohio, 170 million liters of steel-pickling acid and other wastes leaked from hazardous waste injection wells into porous sandstones. The leak was caused by well-casing cracks and corrosion. The responsible company was fined $10 million (Boraiko, 1985).

Although this disposal method is not applicable to solid wastes, it may be possible to inject incinerator ash into deep formations in slurry form. A schematic cross-section of well casing for deep injection of wastes is shown in Figure 8.4.

Injection with Hydrofracturing

A variation of the deep-injection disposal method involves injection of a liquid waste in cement slurry form into the subsurface under such pressure that the liquid fractures or parts the strata and

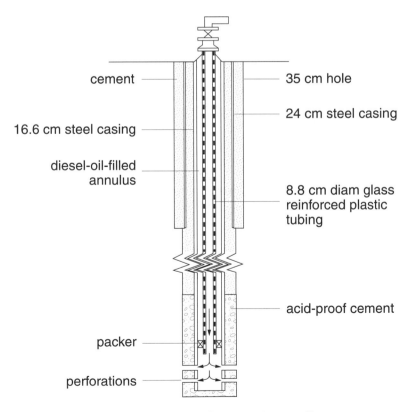

8.4   Schematic cross-section of a deep-injection well.

flows into it. The waste is mixed with cement or other additives so that it will harden once in place. The waste is incorporated into the geologic formation and effectively contained by the weight of the overburden. This method has recently been shown to be feasible and economically competitive under appropriate conditions (Dusseault, 1995). In one case, over 9,000 cubic meters of fine-grained sand coated with heavy oil was injected into a horizontal bed of sand at a depth of 690 meters. Experts believe that this technique could be used for wastes that are inert (i.e., that will not decompose or react with the surrounding strata) and fine-grained or granular, such as ash from incinerators.

The ideal location for injection with hydrofracturing would have the following characteristics: geology consisting of flat-lying sediments; a target horizon that is permeable and porous; a target horizon hydrologically isolated from aquifers that contain potentially usable water, and from the ground surface, by thick shale and clay layers; and no nearby faults, oil reservoirs, soluble salts, or potentially valuable resources. Many sites in sedimentary basins meet these specifications.

Exotic Solutions

Several imaginative methods have been proposed to deal with wastes, especially nuclear wastes. The goal is to place these substances where they cannot impinge on living organisms.

Polar Ice Cap Disposal

A proposal has been put forward to dispose nuclear wastes in the polar ice caps. The rationale is that very few humans or other living beings would come in contact with the waste in these isolated regions. Furthermore, if the wastes were frozen into the ice, they should remain effectively in a deep freeze forever.

Needless to say, this disposal method faces insurmountable logistical complications, such as costs, difficulty of transportation, and access to international territory. Furthermore, the future effects of global warming need to be considered. Like dumping in the oceans, this approach seems attractive primarily because of the remoteness of the area. There is, however, potential for polluting an otherwise pristine area whose environmental sensitivity is not fully understood.

Sub-Seabed Disposal

A variation of ocean dumping has been proposed for nuclear wastes, although it has never been implemented. This method involves encapsulating waste in cylinders which are shaped somewhat like rockets with pointed noses. The cylinders would be dropped from a ship into the ocean and would fall to the bottom and burrow a few meters into the ooze on the sea floor. Ongoing sedimentation would continue to add cover over the tops of the cylinders, albeit at a slow rate. Thus, the wastes would become sealed below the seabed, an area that is isolated and would prevent or greatly slow leaching and escape of the contents. Even if some radionuclides were to escape, they would be greatly diluted by sea water.

The cost of manufacturing appropriate cylinders makes this method prohibitively expensive for all but the most concentrated wastes. There are also problems related to the international status of the oceans and the various anti-dumping conventions in force at this time.

An interesting variation of seabed disposal is a proposal to drop the wastes into deep ocean trenches where tectonic plates bump into each other and one plate submerges into the Earth. Not only are these the deepest parts of the ocean; in addition, the subducting plate would, over millions of years, carry the waste tens to hundreds of kilometers below the Earth's surface. Such ocean trenches occur in a long arc along the western coast of South and Central America, and in another long line stretching from the Aleutian Islands to the eastern coast of Japan to north of New Zealand. A potential problem is the fact that the friction between the two plates causes enormous heat, so that the islands and continental coasts next to ocean trenches have abundant volcanoes. It might cause future generations consternation to see volcanoes spewing up their ancestors' garbage.

Shooting Wastes into Outer Space

One proposal that has been put forth is truly in the realm of science fiction: wastes would be propelled into outer space. In theory, this is a great idea. Waste shot into the Sun would be completely incinerated and destroyed. However, the cost of rocket launches is currently prohibitive, even for the most toxic of wastes. Furthermore, should an accident happen during the launch—as in the disastrous explosion of the space shuttle *Challenger* in January 1986—toxic contaminants would be sprayed over large tracts of land and water.

Summary

This chapter has explored a wide range of alternatives for the disposal of solid waste, including waste emplacement under the oceans, deep underground, into polar icecaps, or even into outer space. Of the disposal alternatives discussed, two in particular have potential. The first is disposal deep underground. Solid rock offers considerable containment and protection. It makes sense to place wastes deep underground where they will be contained and isolated, rather than on the surface where erosion and weathering—not to mention human curiosity—can attack and deteriorate engineered and natural barriers. Although the human race has explored and sent probes millions of kilometers into outer space, our exploration of the planet's inner space, is limited to shallow probings by boreholes and mines. Thus, the disposal of wastes, of whatever kind, will be considerably safer deep underground in carefully selected locations, than in surface landfills near urban centers. Containment is provided by geological formations that have been stable for many tens or even hundreds of millions of years. It can be predicted with some confidence that they will retain this stability for eons to come.

Underground disposal facilities should be located in sites with the following characteristics:

- Structurally stable rock
- Little or no economic value
- Groundwater in small quantities and moving slowly; preferably saline or unusable

An advantage of deep underground disposal is that it does not utilize space on the ground surface, which remains free for other uses. It would also be very easy to collect methane from underground chambers containing municipal waste.

Attractive as underground disposal appears, it is economically prohibitive for municipal solid wastes at this time owing to their large volumes, although it may become feasible once a greater diversion of municipal wastes is achieved. This method has already been shown to be feasible in Germany for hazardous wastes.

The second alternative that deserves serious consideration—and is starting to receive some use—is landfill mining in conjunction with incineration and energy recovery. This can extend the lifespan of landfills manyfold, as well as allowing engineering upgrades to be made to older landfills.

Discussion Topics and Assignments

1. With one or two friends or classmates brainstorm how to dispose of solid wastes in ingenious ways that have not generally received consideration. How many ways can you devise for safe disposal if cost is no obstacle? What are their pros and cons? Can you think of any methods that are economically feasible?

2. What do you think of ocean disposal? Should it be done at all? For some wastes? For all wastes? If it were to proceed on some limited basis, how should disposal in international waters be regulated, monitored, and enforced?

Suggested Reading

Cousteau, Jacques, and D. James. 1988. *The Living Sea.* New York: Nick Lyons.

Duxbury, Alison, and Alyn Duxbury. 1984. *An Introduction to the World Oceans.* Dubuque, Iowa: Wm. C. Brown.

Greenpeace. 1987. *Coastline: Britain's Threatened Heritage.* London: Kingfisher.

Gregory, C. E. 1983. *Rudiments of Mining Practice.* Houston, Tex.: Gulf.

Kullenberg, G. (ed.). 1986. *The Role of the Oceans as a Waste Disposal Option,:Proceedings of a NATO Conference.* Boston: D. Reidel.

Vesilind, P. J. 1989. The Baltic: Arena of Power. *National Geographic,* May, pp. 602–635.

9

INCINERATION
The Burning Issue

Fire has always held a fascination for humans, and it has been one of our most useful tools. Fire has provided warmth, cooked food, cleared forest lands, offered protection against marauding animals, and much more. Although garbage has probably been burned ever since humans discovered fire, it has been incinerated in a systematic manner for only about a century. Perhaps surprisingly, given its long history and obvious benefits, waste incineration is a topic that is both controversial and emotional. In this chapter we will discuss the advantages and disadvantages of incineration and how it can contribute to an integrated waste management program.

Under proper conditions, incineration provides a number of benefits:

- It greatly reduces the volume of waste that must go to disposal in landfills—a vitally important objective. In conventional municipal incinerators, the volume reduction ranges from 80% to 95%, with a mean of about 90%.
- It can be used in conjunction with landfill mining (see chapter 8) to reclaim closed landfills and greatly extend the operating lifetimes of existing landfills.
- The ash produced is relatively homogeneous and thus more suitable than raw waste for treatment such as solidification in concrete.
- A relatively large proportion of the organic compounds, including putrescible and hazardous wastes, is destroyed; thus, there is a net reduction in the quantity of toxics.

- Energy can be generated as a useful byproduct, which preserves nonrenewable fuels like natural gas, oil, and coal. Fewer air pollutants are produced by burning waste than by burning coal or oil.

The use of incineration has been increasing in the United States since about the mid-1980s, and currently the country burns about 16% of its municipal wastes (EPA, 1994). This figure is significantly lower in Canada—about 4%—but it can be much higher overseas. For example, Japan, which faced its waste disposal crisis in the 1950s, 20 years before the crisis reached North America, incinerates approximately 34% of its municipal garbage (Hershkowitz & Salerni, 1987). Most Japanese incinerators generate electricity. In Sweden, the government regards waste as a resource, not something to be squandered by landfilling; approximately 41% of its waste is incinerated in 21 waste-to-energy incinerators, with almost all the energy being delivered to district heating systems (Rylander, 1994). This energy corresponds to 4.5 terawatt-hours (tera means 10 raised to the power 12), or 15% of the total district heating requirements in Sweden. There are more than 400 waste incinerators in the world.

The main drawback to incineration is that the process releases contaminants into the air, violating the principle of protecting health and environment. Thus, if incineration is to be used, it must incorporate rigorous emission controls. There is considerable opposition by the public to the use of waste incinerators, at least partly because older incinerators certainly caused air pollution. Modern waste-to-energy plants have largely overcome this deficiency by including improved combustion processes, better pollution control technology, and the production of a useful product, energy.

Opponents of incineration argue that contaminants are spread into the atmosphere where they cannot be controlled, instead of being contained in a landfill. Another disadvantage of an incinerator is that it is more costly to construct than a landfill; furthermore, all of the capital cost is incurred up front, whereas landfill capital costs are spread over the operating lifetime. Incinerator technology is far more sophisticated than that of a landfill, requiring more careful control and trained operators.

Design criteria for incinerators should ensure that:

- Air will be supplied in the quantities needed for proper combustion.

- Air, waste, and combustion products will be mixed to ensure complete combustion.
- Gases will be tempered and cooled to prevent damage to the refractories (heat-resistant incinerator liner) and to allow the gases to be treated.
- Particulates and noxious substances will be removed from the flue gases.
- Waste will be fed into the furnace and ash removed without allowing combustion products to escape.
- A water treatment plant will be incorporated to process the water used in cooling the ash residues and flue gases.
- The amount of maintenance and downtime for repairs will be minimized.

Types of Incinerators

Three standard and two less common models of incinerator are used in North America. Each can be operated, with some modifications, to produce energy.

The *mass-burn incinerator* is the most common type and is similar to a coal-fired steam boiler. A schematic cross section of a mass-burn incinerator is shown in Figure 9.1, and an aerial view of an actual incinerator in Figure 11.26. An advantage of mass-burn incinerators is that the waste requires minimal processing. Mixed

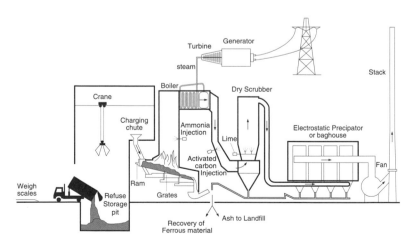

9.1   Schematic view of a mass-burn waste-to-energy incinerator.

garbage, from which only the largest items such as appliances and logs are removed, is brought to the plant and placed in a large waste storage pit. An overhead crane mixes the refuse to provide a relatively uniform fuel and then loads it into hoppers which carry the waste onto grates in the furnace. Fans in the furnace floor and walls provide air for the oxidation (i.e., combustion) process. The waste is burned at an optimal temperature of about 1100°C and remains on the grate for 45 to 70 minutes to ensure complete combustion. The gases that form are heated by supplemental fuel injection for an additional second or two to ensure complete destruction of resistant chemicals. The hot gases are then cooled by water in boiler tubes that generate steam for electricity, heating, or other purposes. Then the gases are sent to pollution control devices, which may include ammonia injection for NO_x (nitrogen oxides) control, a dry scrubber for SO_2 and acid gas control, carbon injection to remove mercury and dioxin, and a baghouse to remove particulate matter.

The ash that accumulates at the bottom of the furnace is removed through a water-quenched conveyor and emptied into a storage area from which it is periodically removed and transported to a landfill. Some plants remove and recycle the larger pieces of iron and other metals that have not burned. Fly ash is collected from the dry scrubber and baghouse and taken to a landfill. Mass-burn incinerators can have capacities of 90 to 2,700 tonnes of garbage per day. A case history of a mass-burn incinerator is presented in chapter 11.

A *modular incinerator* is similar to a mass-burn incinerator but typically has a smaller capacity, in the range of 14 to 365 tonnes of waste per day. It is modular in design and can be built in units at the factory and then shipped to the facility site.

A *refuse-derived fuel (RDF) incinerator* burns garbage that has been processed before being burned. Although processing is required, the prepared fuel will be consistent and will meet specifications for energy content, moisture, and ash content. A significant advantage is that recyclable materials such as iron, aluminum, and glass can be removed during the processing. The RDF can be produced in shredded or fluff form, or it can be compacted into a denser fuel such as pellets or cubes. Densified RDF is more costly to produce, but it has the advantage of being easier to transport and store. This fuel works most effectively in specially designed boilers, but it can also be used in coal-fired boilers. RDF has an energy value comparable to that of coal and can be used either alone or mixed with coal. Because of the higher energy content and more uniform nature of the fuel, RDF incinerators are smaller and can be more effectively controlled than mass-burn units of similar capacity. By the end of 1992, RDF facilities accounted for about 20% of the waste-to-energy plants in the United States.

The *fluidized-bed incinerator* is a relatively new technology in North America for garbage, although it has been used to burn sludges. This incinerator injects refuse-derived fuel into a loose, moving bed of limestone and sand, which is suspended above the furnace floor, like a fluid, by an upward flow of air. The "fluidized" bed of sand and limestone helps to distribute the heat evenly throughout the burn, resulting in more complete combustion and more uniform ash quality. In addition, the limestone helps to neutralize acids; this, combined with the higher combustion efficiency, results in lower emissions of nitrogen oxides, sulphur dioxide, and dioxins than occurs from the other types of incinerators.

Because fluidized-bed incinerators require preprocessing of waste, they fit well with materials recycling. These incinerators, being much smaller than mass-burn incinerators, may be more appropriate for smaller communities.

Rotary kiln furnaces similar to those used in the cement industry can be used for incinerating wastes. The kilns are large, gently sloped cylinders lined with refractory (heat-resistant) materials that rotate slowly while being heated to very high temperatures. A supplementary fuel such as oil or gas is generally used. The kiln system is very flexible and can handle a wide variety of waste types and sizes. Kilns 2.5 or 3.0 meters in diameter are common and can handle large waste pieces, including drums. The kilns are slightly inclined so that waste moves down the slope. The length of the kiln and the amount of incline control the time of exposure of the waste to high temperatures, and these features can be designed to provide the required destruction level. Some kilns are designed to maintain a layer of melted glasslike slag on the inside of the drum; this protects the lining, or refractory, of the furnace from the high temperatures and prolongs its life; it also produces a more leach-resistant vitrified ash residue and helps to capture fine combustion particles. Gas scrubbers and dust removal systems are easily attached.

In all incinerators, the hot gases produced by incineration must be cooled to stop chemical reactions and to protect the downstream pollution-control equipment. Cooling is usually done by quenching the hot gases with large volumes of water. The water and condensate are sent to a wastewater treatment plant, which forms a necessary part of a modern incinerator.

Waste to Fuel Cement Kilns?

Instead of building new incinerators, waste can be used to replace fossil fuels that are burned in existing rotary kiln furnaces used to make cement. Because of the requirements of the cement-making process, these furnaces operate at the same high temperatures and long residence times that are required to incinerate wastes. When properly managed, this is a win-win situation. The cement industry benefits because wastes replace expensive fossil fuels. The environment benefits because wastes are destroyed that would otherwise be placed in landfill; in addition, renewable resources are saved. In the United States and Canada, about 9% and 3%, respectively, of the fuel used in cement plants are provided by waste. The figure is as high as 25% in some European countries. Wastes commonly used in cement kilns include scrap tires, spent solvents, used oil, and wood wastes, all of which have high thermal content. Municipal refuse and refuse-derived fuels could also be used.

By the early 1990s there were 160 municipal waste incinerators operating in the United States; Table 9.1 lists them by type. Mass-burn incinerators were the predominant type, and the majority created electrical energy. In addition, there were approximately 6,000 medical incinerators operating mostly at hospitals, attesting to the ability of combustion to destroy germs and pathogens.

An incinerator's efficiency of combustion is an important factor. As combustion efficiency increases, more organic materials are destroyed and less dioxin and other pollutants are formed. Combustion efficiency is determined by temperature, combustion time, turbulence (which allows mixing of the fuel with oxygen), and uniform conditions throughout the burn volume. Typical mass-burn incin-

Table 9.1. Municipal incinerators operating in the United States in 1992 (Murphy, 1993).

Technology	No. Facilities	Capacity (tonnes/day)
Incinerator	34	6,330
Mass-burn waste-to-energy incinerator	65	59,570
Modular waste-to-energy incinerator	48	4,800
Refuse-derived fuels plus other fuel combustion	13	4,680
Total	160	75,380

erator temperatures are greater than 1000°C, and the residence time of waste in the furnace is generally from 45 to 70 minutes. For fluidized-bed incinerators as well as for shredded waste, the residence time can be substantially less.

A Vest-Pocket Incinerator

At the world's northernmost permanently inhabited settlement, they had a garbage problem. The shallow soils, permafrost, and delicate Arctic ecology at Alert, a Canadian armed forces base of about 200 people, were not conducive to building a landfill. The base is accessible only by air, and it was very expensive—not to mention messy—to fly Alert's waste to a more southerly landfill. The solution was a small incinerator.

A two-tonne-per-day system was designed and supplied by ECO WASTE Solutions of Burlington, Ontario. Flown up to Alert in two Hercules aircraft, the system was installed and operating in April 1994. Figure 9.2 shows the incinerator being unloaded from a Hercules aircraft, and Figure 9.3 shows it set up. The incinerator operates in a batch mode, with a main chamber where baled waste is burned using oil (or natural gas) at temperatures between 500° and 700°C for about 10 hours in starved-air conditions. The main chamber is cubical and about the size of a large closet (2 meters on a side). Gases and particulates pass into an adjoining cylindrical afterburner where they are further incinerated at about 1000°C in a turbulent environment.

An auxiliary system allows liquids to be injected into the primary chamber so that waste oils and petroleum products, a major waste problem at northern bases, can be destroyed. The unit is automated with computer controls, eliminating the need for a permanent operator on site. After three years of successful operation at Alert, the Department of National Defence is studying ways to use the system to clean up the now defunct DEW Line early warning radar stations that dot the Arctic—sites that would not be an environmental problem today if such incinerators had been installed originally.

This small-incinerator approach holds great promise for many small-volume waste generators such as isolated communities, large ships, and factories; it is already starting to be used by hospitals. Because of the system's inherent simplicity and modular design, it is easy to build larger units or incorporate additional pollution control devices (a wet-scrubber is available), and to extract and use the heat that is generated. More important, because of its small size, relative simplicity, and good environmental performance, it is easy for people to understand this incinerator and the benefits it can bring to their community.

9.2   Incinerator being unloaded in the Arctic from a Hercules aircraft (courtesy of ECO Waste Solutions).

9.3   Incinerator installed in Burlington, Ontario (courtesy of ECO Waste Solutions).

There are many environmental concerns that must be addressed in deciding whether to include an incinerator as part of a waste management system. The two main issues are air emissions and the ash that is produced. Other factors include increased truck traffic and the capital cost of the facility.

Air Emissions

Because an incinerator is based on the principle of combustion, the generation of air emissions is unavoidable. Thus, a critical requirement of modern incinerators is that they have good emission control equipment and meet all applicable air pollution regulations. Indeed, the control of air pollutants is a very difficult and expensive engineering problem that accounts for much of the complexity and cost of a modern waste incinerator.

Health risks associated with these emissions are caused, in particular, by toxic organic compounds such as dioxins and furans, which were first discovered in incinerator emissions in 1977, and by heavy metals such as lead, cadmium (from plastics), and mercury (from batteries). In addition, particulate matter, carbon monoxide, nitrogen oxides, and acid gases such as sulphur dioxide and hydrogen chloride are generated.

Materials that contribute to these harmful emissions include household batteries, lead-acid vehicle batteries, electronic components, and some plastics such as polyvinyl chloride (PVC) that contain lead and cadmium. Yard wastes and other materials with a high moisture content can cause the furnace to burn inconsistently and incompletely, resulting in the formation of dioxins. The following pollutants would be discharged to the atmosphere from incinerating one tonne of solid waste if no pollution control equipment were used (Sarofim, 1977):

Particulates	15 kg.
Sulphur oxides	0.8 kg.
Carbon monoxide	18 kg.
Hydrocarbons	0.8 kg.
Nitrogen oxides	1 kg.
Hydrogen chloride	0.3 kg.

Note that emission control devices remove significant proportions (90% to 99%) of some of these pollutants.

The recognition that incinerator emissions have an adverse impact on human health and the environment, coupled with the passing of the Clean Air Act of 1967 in the United States, resulted in significant improvements in air pollution technology. The U.S. EPA estimates that 95% to 99% of particulate and organic pollutants can be removed from air emissions if appropriate pollution prevention measures are taken. Modern pollution-prevention measures include temperature control, scrubbers, baghouses, and precipitators, described below.

The incinerator should be operated at a constant and uniform temperature. When combustion temperatures are below about 800°C, odorous emissions may occur. Combustion temperatures of greater than 1000°C minimize the emission of carbon monoxide, dioxins, furans, volatile organic compounds (VOCs), and other potentially hazardous compounds in the flue gas. Combustion control appears to be one of the best ways of reducing emissions of dioxins and furans. Since the formation of carbon monoxide (CO) is a sign of incomplete combustion, it can be used as a "surrogate" to monitor dioxins and furans. This is very useful because CO can be measured using continuous emission monitoring methods, whereas dioxins and furans cannot.

Scrubbers are used to control acid gases and to cool the flue gas, causing pollutants to condense before the gas enters the next stage of the pollution control system. Either wet or dry scrubbers can be used. Dry scrubbers, also known as dry sorbent injection scrubbers, immerse the emissions in a very fine powder lime which neutralizes the acid gases and also improves mercury capture. The dry scrubber system has a low capital cost and can be installed easily in existing facilities. Wet scrubbers use the same principle as dry scrubbers, but they spray a lime-and-water mixture onto the emissions to convert the gases into liquids and solids for collection in an electrostatic precipitator. Wet scrubbers can be placed after, but not before, a baghouse system.

Baghouses are filtering devices composed of a series of large, porous bags through which gases, but not particulates, can pass. A baghouse may contain several hundred individual bags which function on the same principle as vacuum-cleaner bags. Baghouses are replacing electrostatic precipitators as the pollution control technology of choice because they collect more particulate matter and are better at trapping very fine particles. This is important because heavy metals and organics tend to attach to smaller particles. Baghouses are used with dry scrubbers.

Electrostatic precipitators remove particulate matter by negatively charging particulates as they pass through the unit and collecting them on positively charged metal plates as they exit the unit. These units can be used with both dry and wet scrubbers.

Prior to about 1967, waste incinerators had few pollution control devices; most of these older incinerators were removed from service following the passage of the Clean Air Act. The energy crisis of the mid-1970s led to renewed interest in waste-to-energy incinerators, and those built in the 1970s and early 1980s employed baghouses or electrostatic precipitators and were as clean, or cleaner, than most power plants using coal or oil. Almost none of these plants had acid-gas controls. In 1990, the United States made a national commitment to add acid-gas controls to most power plants, including waste-to-energy incinerators. At the same time, many plants were also adding other, more advanced pollution control technologies.

Dry scrubbers, described above, control acid gas emissions. More advanced upgrades include controls for nitrogen oxide and mercury. Nitrogen oxide is composed of NO and NO_2, which are collectively known as NO_x. These are formed as a result of the thermal interaction beween oxygen and nitrogen in air, as well as nitrogen contained in the waste, during the combustion process. Nitrogen oxides are precursors to ozone (O_3) and peroxyacetal nitrate (PAN), the photochemical oxidants that form smog. Nitrogen oxides can be converted to nitrogen and water by spraying ammonia or urea into the hot furnace exhaust. Blowing activated charcoal into the exhaust gas prior to the baghouse is an effective control device for dioxins and furans, mercury, and many semivolatile products of incomplete combustion such as chlorobenzenes, PAHs, and PCBs, as well as particulates.

Table 9.2 summarizes the pollution control devices currently in use and the contaminants they prevent or collect.

In the United States, an acceptable incinerator for hazardous wastes has been defined as one that reduces the amount of any particular toxic compound by 99.99%. This is known as the "four nines" standard. In some cases, most notably for PCB incineration, a destruction level of 99.9999%, or six nines, is required.

Table 9.2   Summary of incinerator pollution control technologies.

Technology	Contaminants Affected	Date Introduced
Good combustion control	CO, dioxins, furans, VOCs, and others	before 1967
Baghouse	Particulates	1967
Electrostatic precipitator	Particulates	1967
Wet/dry scrubber	Acid gases: SO_2, HCl	1990
Ammonia/urea spray	NO_x	late 1990s
Activated charcoal spray	Hg, dioxins, furans, others	late 1990s

Dioxins and Furans

The emission of dioxins and furans has become one of the most complex and controversial issues surrounding waste incineration. Dioxins are a family of organic chemical compounds known as polychlorinated dibenzodioxins (PCDDs), with 75 different forms that are characterized by the placement of one to eight chlorine atoms and their aromatic rings. Tetrachlorodibenzo-p-dioxin (TCDD) is the most widely known and is found as an unwanted contaminant in pesticides, wood preservatives, and defoliants. In particular, its association as a contaminant in the defoliant Agent Orange has helped give dioxin the reputation of one of the most toxic chemicals on Earth.

Furans (polychlorinated dibenzofurans or PCDFs) are a family of 135 organic compounds. The best known is polychlorinated dibenzo-p-furan. They are closely related to dioxins.

Of the total of 210 compounds, seventeen are considered to be particularly harmful to the environment. They are very stable and readily bioaccumulate in fatty tissue. Symptoms of human exposure to dioxin include chloracne (a skin disorder), dizziness, and digestive disorders. In the environment, these compounds are almost always found adsorbed onto particulates such as soil, sediment, and ash. There has been considerable controversy about the health effects of dioxins (Crittenden, 1995). The U.S. EPA conducted an extensive study and concluded that dioxins are "likely to present a cancer threat to humans." Many scientists have discredited the EPA's aggressive regulatory stance against dioxins and claim that the EPA has overstated the risks. Although some dioxin and furan compounds have been shown to be extremely toxic and carcinogenic in animal tests, statistical evidence has been presented that show that dioxins do not produce cancer in humans (Crittenden, 1995).

Although some dioxin is made synthetically, it is also a natural byproduct of most combustion processes and is created by forest fires, woodstoves, automobiles, power plants, metal smelting, and cigarettes. Although incineration has often been represented as a major cause of dioxins, recent EPA data indicate that the incineration of municipal waste accounts for approximately 0.7% of the dioxin in the U.S. environment. This is a small fraction of the dioxin created by woodstoves. Studies in Germany indicate that although some dioxins are formed and some destroyed in waste incineration, the net effect is a decrease in dioxins (Lahl et al., 1990).

There is considerable uncertainty about the exact mechanism of how dioxins and furans are formed in incinerators. A study in which the amounts of chlorine introduced via plastics were reduced, did not show a reduction in

→

> **Dioxins and Furans** (continued)
>
> the amounts of dioxin/furan emissions (although it did show a reduction in the generation of hydrogen chloride). There is a significant correlation between incomplete combustion and formation of dioxin/furan emissions, indicating that good combustion control is the best way to limit their emission. It has also been shown that dioxins are formed when a hot particulate-laden gas is cooled through the 200° to 400°C window slowly (i.e., in a period of seconds), as occurs with heat recovery, electrostatic precipitators, and baghouses. Wet scrubbers, which cool gases through this window in a small fraction of a second, generally have low levels of dioxins in the stack gas.
>
> Dioxins provide a classic example of the sensationalism that often pervades the waste management field. In reality, our understanding of dioxins and furans and their impact is very incomplete owing to the large number of similar compounds, the complex systems within which they exist (biosphere, waste management facilities), and their low concentrations, as well as because analytical techniques for their detection became available only recently. However, this has not prevented both sides from making unsubstantiated and often outrageous claims.

Air emissions created by waste-to-energy (WTE) incinerators can be viewed on a relative rather than absolute scale. In other words, let us compare the emissions from WTE plants with those generated by other types of electrical energy sources. Getz (1994) performed a comparative study of the pollutants emitted by different fossil-fuel power plants, including those fired by coal, natural gas, oil, and waste. Emissions of nitrogen oxides, hydrogen chloride, sulphur dioxide, non-methane hydrocarbons, methane, carbon dioxide, particulates, and some heavy metals were studied. The amounts emitted were calculated on a unit-energy basis (i.e., kilogram of emission per megawatt-hour of power generated).

Based on these comparisons, the researchers concluded that the emissions from natural gas power plants are the cleanest. However, most of the emissions from WTE incinerators are either lower than or essentially equivalent to the emissions from oil and coal power generation facilities on a net electricity production basis. In fact, for many cases, especially for the trace metals, WTE facilities generate emissions that are five to ten times cleaner than those from oil and coal plants.

This is an important finding. It shows that electrical power from waste incineration is cleaner than that from coal or oil power plants. Thus, WTE should be preferred over these two alternatives; it will not only help to preserve these valuable resources for the future, but it will also cause less long-term damage to the environment in the process. In 1991, about 0.2% of all electrical power in the United States was generated by waste incineration, whereas about 58% was produced by oil and coal.

The U.S. government has recognized that burning garbage for energy is an environmentally sound strategy. For example, the U.S. Clean Air Act of 1990 gives credits to utilities for reducing pollution by buying power from waste-to-energy incinerators. The U.S. National Energy Policy Act of 1992 calls for greater use of incineration of municipal waste for energy as a means of reducing greenhouse gas buildup. The U.S. Department of Energy estimates that waste-to-energy technology will be one of the four largest contributors to the nation's planned carbon dioxide reductions for the year 2000, accounting for 15% of the total.

It must be recognized that if wastes (after recycling) are not incinerated, they will be placed in landfills. In addition to the increased air pollution from substituting less cleanly burning fossil fuel for incineration, there are also significant air emissions from landfills. While incineration releases more carbon dioxide, landfills release more methane, which has a global warming potential 20 times greater than that of carbon dioxide.

How Do Landfills and Incinerators Stack Up?

A Royal Commission on Environmental Pollution in Great Britain studied greenhouse emissions from both landfills and incinerators (Royal Commission, 1993). They reported that incinerating 1 million tonnes of municipal garbage produces net emissions of 15,000 tonnes of carbon in the form of carbon dioxide, whereas landfilling it with energy recovery produces emissions of greenhouse gases equivalent to 50,000 tonnes of carbon as carbon dioxide. Thus, even with gas collection and burning, landfills have a worse impact on global warming than incinerators.

Furthermore, it is only recently being recognized that trace gases emitted from landfills can contain compounds—such as vinyl chloride, benzene, and trichloromethane—that are harmful to human health (Birmingham et al., 1996). A complete analysis of incineration should recognize that the alternative to incineration, landfilling, has significant gaseous emissions associated with it. As noted in chapter 7, recent studies indicate that gas emissions from modern waste incinerators pose a lesser health risk than emissions from landfills, even if the landfills have gas control systems.

A sophisticated monitoring system that controls operations and ensures that emissions are being properly abated is an integral part of an incinerator. Typical systems consist of contaminant monitoring at the stack and at other points. Today, waste incinerators generally employ continuous stack emission monitoring for pollutants such as carbon monoxide, sulfur dioxide, and nitric oxides. The data are transmitted directly not only to the on-site control office but also to the regulatory authority.

What to Do with the Ash?

Ash is a significant byproduct of incineration: as much as 20% to 25% by weight of ash residue can be generated. There are two kinds of ash produced, fly ash and bottom ash. The former includes charred paper, cinders, soot, and other light materials that rise and travel with the hot gases and are captured by baghouses or electrostatic precipitators. Although fly ash accounts for only about 10% to 25% of the total ash, it is generally more toxic than bottom ash because heavy metals and dioxin are attracted to and condense on the small fly ash particles.

Bottom ash is composed of noncombustible and incompletely burned refuse that is left in the bottom of the combustion chamber. It comprises 75% to 90% of the total ash by weight. Most facilities combine the fly and bottom ashes and then cool the ash by quenching with water. This prevents the fine particles from blowing during subsequent handling and transportation.

The toxicity or hazard of the ash is dependent on the composition of the waste and on the efficiency of the combustion process. Ash can contain dioxins, which are created during combustion and cooling, and heavy metals, which are contained in the original waste. Heavy metals commonly found in incinerator ash, especially fly ash, include cadmium, lead, mercury, arsenic, beryllium, zinc, and copper.

It is important to manage ash carefully, because the toxins it contains are in a more mobile and bioavailable form than before incineration. In 1994, the EPA ruled that incinerator ash must be treated as hazardous waste unless toxicity tests prove otherwise. The ash is lightweight and composed of small particles that can readily be dispersed in the surrounding environment. Ash can also be leached easily by water to release heavy metals which can pollute nearby surface and ground waters. Some sound management steps include the following:

- Ash should be covered while it is in temporary storage or during transportation.
- Fly ash and bottom ash should be managed separately because the fly ash is more toxic.
- Ash should be stabilized through chemical or physical treatment to improve its leach resistance and to reduce its toxicity.
- Source reduction should be practiced; that is, materials containing heavy metals should be kept from entering the incinerator in the first place. For example, household batteries could be designed to exclude or minimize the use of cadmium. The practice of household hazardous waste collection days should be encouraged.

The amount of ash can be minimized by removing incombustibles from the waste stream it is fed into the incinerator. In Japan, for example, because of the widespread separation of noncombustible material from incinerator-bound waste, the volume of ash residue is less than the 10% to 20% characteristic of American plants (Hershkowitz & Salerni, 1987).

For decades, nearly all fly ash and bottom ash were buried in hazardous waste landfills to minimize the risk they pose. In recent years, however, methods have become available for stabilizing the ash through vitrification and solidification. Vitrification involves melting the ash and then cooling it into a glasslike substance which can be disposed of in blocks or spun into insulating materials. Solidification, the least expensive and simplest stabilization method, involves the addition of an adhesive substance to form a hard, concretelike material. Both methods create an inert, leach-resistant material which essentially locks in the heavy metals and other toxic substances.

It would be environmentally beneficial if safe uses for incinerator ash could be found. One potential use is in road construction. As early as 1975, ash aggregate received approval for use in highway construction in Houston, Texas. The material, called "Littercrete," consists of 89% ash aggregate, 9% asphalt, and 2% lime. Topped with asphalt, the roadbed has exceeded performance standards for conventional road materials over the past 16 years (Woods, 1991).

Research has been conducted on the use of ash in marine piers, seawalls, and other structures. Many European countries permit the use of bottom ash as an aggregate in roadbeds, cinderblocks, and other materials. The United States and Canada lag behind Europe in this regard, and most of their incinerator ash is taken to landfill.

Other Factors

Like other types of waste management facilities, incinerators generate a significant volume of truck traffic. At a large incinerator, several hundred trucks may visit the site each day to deliver garbage and to remove ash to landfills. The impact of the truck traffic needs to be considered in the siting of such a facility, just as it would be for a landfill or material recycling plant.

Economics is a significant factor in considering an incinerator, since it is a capital-intensive facility compared to a landfill. Furthermore, the entire facility must be constructed prior to its operation rather than being developed throughout its lifetime like a landfill. In addition, pollution control technology and ash disposal contribute significantly to the facility's cost.

The Case Against Incineration

In recent years, in spite of the desperate need for alternatives to landfilling, incineration has raised considerable debate, controversy, and opposition. In Ontario, for example, the provincial government opposed waste incineration and banned the construction of new incinerators between 1991 and 1995. Only four incinerators are currently operating in the province.

The main arguments put forth by opponents of incineration are:

- Contaminants are dispersed into the atmosphere, contributing to global warming, acid rain, and respiratory health problems. In other words, incineration merely transfers contaminants from one medium to another—from the ground to the atmosphere.
- A new type of waste, ash, is created that is more toxic and bioavailable than the original waste.
- New, highly toxic compounds—dioxins and furans—are created.

- Incineration needs paper and other combustibles and thus competes with and detracts from recycling programs.

Let us study each of these arguments in turn. While it is true that incinerating waste contributes to atmospheric pollution, it must be realized that the alternative—placing waste in landfills—also contributes significantly to pollution of the atmosphere. In fact, controlling emissions from a landfill is much more difficult, owing to its large size and heterogeneous nature, than controlling emissions from an incinerator where the combustion gases are all funnelled through a stack. Moreover, atmospheric pollution by a landfill continues for many decades after it has closed.

Furthermore, landfills will eventually leak and contaminate the groundwater. How do we compare groundwater contamination from landfilling against air pollution from incineration? If the specific geographic, atmospheric, geologic, and demographic settings were known, detailed risk analyses could be performed and the health impacts of each scenario could be estimated. Even then, comparison is complex: air pollution would lead to more immediate impacts, whereas the effects of groundwater pollution would be delayed, probably by years or even decades.

Such a detailed risk assessment was conducted by the Ontario Ministry of Environment and Energy, comparing a hypothetical municipal landfill that collects and flares gas with a waste incinerator of similar capacity equipped with modern pollution control systems (Birmingham et al., 1996). The study included all possible direct and indirect pathways via air and ground by which nearby residents might receive exposure to contaminants. The results indicated that although both facilities led to acceptable risks, the incinerator was considerably safer than the landfill. A similar study that compared only air emissions from both types of facilities showed that incinerators had a less adverse impact on the environment (Jones, 1994).

In summary, incinerators do transfer pollutants to the air, but they emit less contaminants than if the wastes were landfilled, and they do so under controlled conditions. Furthermore, incinerators prevent groundwater contamination.

Although the generation of ash is often perceived as a negative aspect of incineration, it is also a positive feature. Because the organics have been destroyed, ash will not biodegrade and produce landfill gas and landfill settlement. Because the ash is far more homogeneous than the heterogeneous waste prior to incineration and has greatly reduced volume, it is more amenable to treatment. For example, mixing the ash with lime and water forms a cementlike substance which is hard and relatively leach-resistant. Thus, the ash can be transformed into a material that is more suitable for landfill disposal than the original refuse.

In addition, as discussed earlier, research is being conducted to find safe uses for solidified ash in various materials and construction projects, and a number of successful demonstrations have already been completed. Once economical and safe uses for incinerator ash are fully developed, the amount of waste destined for landfills could be greatly reduced.

In summary, ash is more suitable for landfilling with respect to volume and stability than are unincinerated wastes; it can be improved even more by solidification into concrete. Furthermore, research into beneficial uses for ash will eventually make ash a recyclable commodity.

Although some new toxic compounds such as dioxins are created by incinerators, this is more than offset by the organic compounds that are destroyed by combustion. Furthermore, as discussed in the sidebar, dioxins may not pose as significant a threat to the environment as has been perceived. Nevertheless, dioxin emissions from incinerators should be kept as low as possible.

Incinerators, when properly designed, do not compete with or detract from recycling programs. We saw in chapter 4 that recycling alone is not capable of diverting more than about 50% of waste from landfills. Thus, it is vital that incineration also be used; both are needed, and there is ample "fuel" for both, as has been demonstrated in Japan. Problems would arise only if an incinerator had overcapacity, so that it needed extra fuel to achieve economic efficiency. However, planning the size of an incinerator to complement a vigorous recycling program is an easy task, as has been demonstrated by dozens of communities that have both incinerators and successful recycling programs. A case history illustrating the harmonious coexistence of incineration and recycling is presented in chapter 11.

In summary, incineration, coupled with a strong recycling program, can greatly decrease the need for landfills, and it should form an integral part of any waste management system. The benefits of incineration have long been recognized in Europe and Japan, where the use of incinerators is common. We should learn from their experiences and work to improve the way that incinerators are operated. For example, the Japanese method of presorting the waste to

remove hazardous and noncombustible materials could be introduced; incinerator operators could receive more stringent training;
and the fines levied for any violations or pollution emissions could
be increased. Finally, vigorous research into more advanced emission controls and ash treatment should be encouraged.

Discussion Topics and Assignments

1. Based on the list of waste you created in one week (see
   Question 4 in chapter 4), how much of the waste that is not
   recycled is combustible (i.e., paper, plastic, etc.)? Based on
   your qualitative observations, could a strong recycling program and an incinerator work together in your community?
   What impact would this have on the size of landfill?
2. Using the information presented below, estimate the energy
   content in kilojoules(kJ)/kg of the waste before and after
   recycling. The recycling program removes 40% of paper,
   70% of cardboard, 33% of plastics, 25% of glass, and 22%
   of tin cans.

Waste Item	Weight (kg)	Energy (kJ/kg)
Paper	34.0	16,280
Cardboard	6.0	15,820
Plastics	7.5	33,030
Textiles	2.0	17,440
Rubber	0.5	23,260
Yard wastes	18.5	6,510
Food wastes	9.0	4,880
Wood	2.0	18,610
Glass	8.5	160
Tin cans	6.0	700
Other metal	3.0	700
Dirt, ash, etc.	3.0	6,980
Total	100.0	

If the minimum energy content required for an incinerator
is 9,300 kJ/kg, are the waste streams before and after recycling suitable for incineration?
3. Research and prepare a short essay on dioxins and furans.
4. Suppose you are on a citizens' action group reviewing a
   proposal for a waste incinerator in your community. What
   requirements would you impose before you would accept
   the incinerator?

Suggested Reading

AIMS Coalition. 1994. *Waste-to-Energy: Making a Clean Energy Source Cleaner*. Silver Spring, Md.: Solid Waste Association of North America.

Clarke, M., et al. 1991. *Burning Garbage in the US*. New York: INFORM.

Murphy, Pamela. 1993. *The Garbage Primer: A Handbook for Citizens*. New York: League of Women Voters Education Fund, Lyons & Burford.

Tchobanoglous, G., H. Thiesen, and S. Vigil. 1993. *Integrated Solid Waste Management: Engineering Principles and Management Issues*. New York: McGraw-Hill.

Tillman, D. 1991. *The Combustion of Solid Fuels and Waste*. San Diego, Calif.: Academic Press.

10

CONTAINMENT, ENCAPSULATION, AND TREATMENT

Just as a chain is only as strong as its weakest link, a landfill will only function as well as its weakest component. The most important "links" of a landfill are the cover and bottom liner that provide watertightness. Because of their critical significance, this chapter is devoted to studying the materials from which these barriers are constructed and how they are emplaced. We also look at ways in which the wastes themselves can be converted to forms that are more suitable for long-term disposal.

Plastics and Polymers

Polymeric membranes, more commonly known as geomembranes or flexible membrane liners, are widely used in both the cover and the bottom liner systems. These synthetic materials have gained acceptance as barriers at landfills because they exhibit very low permeabilities, they are resistant to many chemicals, and they can often be installed for less cost than comparable clay liners.

The polymeric membranes used in landfills consist of synthetic plastic or rubber sheets that are joined together in the field using solvents, adhesives, or welding processes to form continuous liners. There are several polymers and compounds that are used, and these have a wide range of material properties. The most common materials are high-density polyethylene (HDPE), polyvinyl chloride (PVC), chlorinated polyethylene, and butyl rubber. Of these, HDPE is most commonly used for landfills, in part because of its documented resistance to a wide range of chemicals. Its thickness ranges from about 0.75 to 3 millimeters (30 to 120 mils).

Many polymers can be made in either vulcanized form (treated with sulphur and heat to give strength by building crosslinks between the rubber polymer molecules) or unvulcanized (thermoplastic) form. Vulcanized liners tend to be stronger and more chemically resistant, but thermoplastic versions of compounds like chlorinated polyethylene are more commonly used because they are easier to seam and repair in the field. A number of additives can be introduced to improve characteristics such as stiffness or flexibility, resistance to fungicides and biocides, and resistance to ultraviolet light and ozone.

Nomenclature of Plastics Used in Landfills

A whole new family of synthetic materials has come into use in the past decade or two. Known as geosynthetics, they are used in a wide variety of civil, geotechnical, and environmental engineering applications—constructing slopes, earthen dams, lagoons, landfills, and much more. Here is a brief explanation of the main types of materials.

Geomembranes are thin sheets of impervious rubber or plastic, used as liquid or vapor barriers.

Geotextiles are similar to regular textiles but consist of synthetic fibers rather than natural ones. They are flexible, porous fabrics, generally used to separate materials of different sizes—for example to prevent a sand layer from leaking into a stone layer—while allowing gases and fluids to pass. They can also be used for reinforcement.

Geogrids are plastics with a very open, strong gridlike structure with large apertures. They are used to provide reinforcement.

Geonets consist of parallel sets of polymeric ribs that allow liquid to flow within the open spaces. They are used to provide drainage for fluids and can save substantial space compared to soil drainage layers. They are always overlain by a geotextile to prevent clogging by overlying soil.

Polymeric liners (geomembranes) are generally considered to be virtually impermeable if they remain intact. However, they can have pinholes caused by grit in the manufacturing process; they are susceptible to punctures and tears; and they can have faulty seams if rigorous quality control is not exercised. Problems that can arise during placement include dropping tools on the liner, driving trucks on the unprotected liner, high winds getting under the liner, use of excessive force in moving and placing the liner, and incomplete or faulty welding.

Landfill construction requires that large rolls or panels of geomembrane liners, usually 4.5 meters (15 feet) wide, be joined in the field. This is a critically important operation that requires careful attention so that the seams do not have zones of weakness or leakage.

There are four general categories of seaming methods: extrusion welding, thermal fusion, solvent fusion, and adhesive seaming. Only the first two are applicable to high-density polyethylene (HDPE), which is most commonly used for landfill liners. In extrusion welding, a ribbon of molten polymer is extruded over the edge of or in between the two surfaces to be joined. The molten polymer causes the surface of the sheets to become hot and melt, after which the entire mass cools and bonds together. Different seam-welding methods are shown in Figure 10.1.

The hot-wedge seaming method uses an electrically heated resistance element in the shape of a wedge which travels between the two sheets. Roller pressure is applied as the two sheets converge at the tip of the wedge to form the final seam. A hot wedge can make a single seam, while a dual (or split) hot wedge can make two parallel

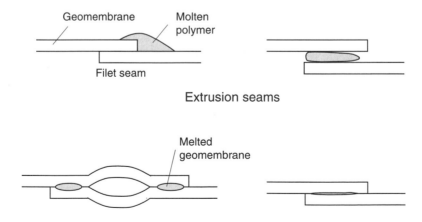

Extrusion seams

Hot wedge and hot air seams

10.1   Methods for welding geomembrane seams.

seams with a uniform unbonded space between them. The latter method has gained popularity because it allows quality control to be conducted by pressurizing the space and then monitoring for pressure drops that might signify leaks. The hot-air method is similar to the hot-wedge method, except that hot air is blown between the geomembrane sheets to cause the melting. A hot-air welding machine is shown in Figure 10.2.

Filet seaming is often the only feasible method for use in poorly accessible areas such as sump bottoms and around pipes and sampling wells. Temperature and seaming rate are important considerations, since too much melting weakens the geomembrane and too little melting results in a seam of poor quality.

Since the seams are vital to the performance of the liner, a number of methods have been developed to test their quality. The vacuum chamber, or box, method uses a box with a window in the top which is placed over the seam, on which a soapy solution has been placed. Then a vacuum is created in the box. If the seam is defective, air enters the box from beneath the geomembrane and bubbles appear in the soapy solution. This method is effective, but progress is slow and the cost is relatively high.

10.2   Hot-air welding of a geomembrane seam in the field (courtesy of SOLMAX Geosynthetics, Inc.).

A more recent method is an electric leak detection system in which voltage is created between one electrode in water over the top of the liner and a grounded electrode beneath the liner. Since geomembranes are insulators, electric current will flow only if there is a leak. Using this method, Darilek et al. (1995) showed in a field test that double fusion welds are superior to single extrusion welds.

The main long-term threats to polymeric membranes are penetration by roots and burrowing animals, tearing or puncturing due to differential settlement of the landfill mass, chemical and microbial attack, and loss of strength and polymer breakdown over time.

There is no assurance that the effective lifespan of polymeric membranes can be much more than about 50 years, and some authors have suggested a useful lifetime of only 25 years. One noted expert, however, feels that geomembranes can have "extremely long service lifetimes" if they are put in place properly (Koerner, 1994, p. 473). Nevertheless, it would be imprudent to rely only on polymeric membranes as the primary long-term barrier in a landfill capping or liner system.

The ability to estimate leakage rates through liner systems is essential not only for environmental impact assessments, but also for designing leachate removal systems. For clay liners a uniform flow is generally applicable, and the standard Darcy Law flow equations can be used. In contrast, geomembrane liners are essentially impermeable and any leakage is controlled by imperfections, usually in the approximate shape of circular holes or linear tears. A variety of mathematical methods have been developed for calculating flow rates through such imperfections (for example, see Oweis and Khera, 1998).

Clay Materials

Both the bottom and top liners in many landfills are constructed of natural soil, such as clay or till with a high clay content, for three reasons. First, suitable low-permeability natural soils are often available at the site or nearby. Second, because clay is a natural geologic material that may have been undisturbed in situ for tens of thousands to millions of years, it is felt that it will also have a long effective lifetime as a landfill barrier. Third, the clay minerals have the capacity to adsorb—that is, capture—various contaminants, thus reducing the toxicity of any leachate that might escape.

Natural clay-containing soils are suitable as liners owing to their intrinsic low hydraulic conductivity. In-situ soil at a landfill site may require reworking and compaction because it has gaps formed by fractures, roots, and animal burrows. If the natural soil is not suitable, other soils may need to be imported; alternatively, materials such as bentonite (a pure clay mineral) may be mixed with the existing soils.

There are three main clay mineral groups: kaolinite, illite, and montmorillonite. Bentonites are clay minerals belonging to the montmorillonite group. Water is readily absorbed between the layers of montmorillonite minerals, causing swelling, which in turn decreases the hydraulic conductivity of the material. One problem with clay liners is the incompatibility of bentonite with leachate. Depending on the leachate chemistry, there may be a tendency for the calcium ion to substitute for the sodium ion in the montmorillonite structure, which causes shrinkage and cracking. An understanding of the chemical and physical properties of these materials is essential in designing an effective liner system.

The chemical interactions between a soil liner and leachate can be complicated. For example, a number of studies have shown that the hydraulic conductivity of the liner can decrease as a result of precipitate formation and the development of active biomass near the upper surface of the liner, which clogs the pores (McBean et al., 1995). In contrast, concentrated organic liquids such as benzene, xylene, or carbon tetrachloride can alter the structure of clay soils and increase their ability to conduct fluids by factors of 100 to 1,000.

A sound monitoring and quality assurance/quality control program is essential for clay liner construction. Important factors include moisture content, the type of compaction machinery, size of clods, bonding between lifts, and uniformity of soil. Additional complications may arise when large expanses of clay liner remain exposed to the elements for extended periods prior to placement of waste.

The main factors affecting the performance of clay liners are hydraulic conductivity, strength, and potential for shrinkage. These three components are related both directly and indirectly to compaction (Oweis and Khera, 1998). Thus, a sound compaction program, complete with quality control, is essential to constructing a clay liner.

Clay barriers can be built which have very low hydraulic conductivity. Field tests have shown that clay-soil liners with an effective hydraulic conductivity of 10^{-8} cm/sec or less are achievable with the use of strict design, installation, and quality assurance guidelines (McBean et al., 1995).However, the following processes can impair the effectiveness of the barrier with time:

- Formation of cracks as a result of seasonal drying and freeze/thaw processes
- Penetration of the barrier by roots and burrowing animals
- Differential settlement of waste, causing cracks to form
- Gas penetration
- Changes in pore water chemistry and ionic composition

Concerns about using clay as a barrier have arisen for the following reasons: there have been cases of liner failures; it is difficult to estimate hydraulic conductivity in the field; and it is felt that leachate chemistry will deteriorate the long-term integrity of the liner.

An interesting recent development is the marriage of clay and synthetic materials into what is called a "geosynthetic clay liner" (GCL). This typically consists of a relatively thin clay layer (usually 4 to 6 mm of bentonite) either sandwiched between two geotextiles or bonded by an adhesive to a geomembrane. The advantages of this material are several: it offers good impermeability; there is good quality control since it is a manufactured product; it occupies much less volume in the landfill, compared to a 0.5 to 1 meter thick clay liner; and because of its thinness, it is easy to install. On the other hand, GCLs have low shear strength and can be used only on relatively flat areas (less than 5% slope); in addition, considerable care is required in installation to avoid damage by construction equipment.

Waste Treatment

Waste treatment is defined by the World Health Organization as a chemical, physical, or biological process that reduces the hazard, toxicity, mass, or volume of a waste and makes it more amenable to recycling, further processing, or final disposal.

Another definition might be that treatment is a way of processing waste that in some way helps to achieve one or more of the three basic principles outlined in chapter 2. That is, treatment is any process that:

- Reduces the risk of disposal to human health and the environment
- Reduces the burden on future generations
- Helps conserve resources

For example, improving the quality of the waste for disposal by making it more leach-resistant satisfies the health-and-environment principle as well as contributing to the future generations principle.

There are three main objectives in performing waste treatment:

- Detoxification: either changing the chemistry or destroying certain chemical compounds, particularly those that can harm the environment
- Volume reduction: less wastes need to go to disposal, thus reducing reliance on landfills
- Isolation of waste materials from the environment: improving the way that wastes are encapsulated so that the likelihood of their escaping from the disposal facility is reduced

Two forms of waste treatment are discussed earlier in this book—composting, which uses biological processes to convert waste into a useful form, and incineration, which uses the chemical process oxidation to reduce waste volume.

Waste treatment is a growing field and there are many processes, with more being constantly developed. Most of these, such as neutralization, oxidation-reduction, hydrolysis, reverse osmosis, and ultrafiltration, treat liquid wastes. For solid wastes, the treatment methods are more limited and fall into the following three categories:

Physical Treatment

Various crushing and grinding operations can be used to reduce void spaces and thus the volume of waste that goes into a landfill. This also reduces future problems with settlement, subsidence, and possible cracking of the landfill cover. Tire-shredding and wood-chipping are examples of physical treatment processes. Crushing and grinding, like mulching wood, can also be used to increase the surface area of organic materials to speed up the biodegradation processes in the landfill or in composting. These physical processes are relatively straightforward mechanical operations.

Thermal Treatment

Incineration has become a favored choice for dealing with both municipal and hazardous wastes because most organic compounds can be efficiently destroyed and good volume reduction is achieved. The topic of incineration was dealt with in chapter 9.

Fixation, Encapsulation, and Engineered Barriers

The process of fixation is relatively new, and its application to containing leachable solid wastes is continuing to undergo development. Many of the methods for waste fixation and encapsulation have been pioneered by the nuclear industry. An important potential application in the realm of municipal solid waste is the fixation of incinerator ash.

Encapsulating wastes in a material such as cement prior to placing them in an engineered landfill is akin to placing the contaminants in a box within a box. The process significantly increases leach-resistance, and this can prevent or greatly delay the wastes from being mobilized and entering the groundwater.

Although encapsulation is not economically viable for municipal wastes because of their great volume and heterogeneity, this method is used for specialized wastes such as low-level radioactive wastes and hazardous wastes that occur in smaller quantities. Cement, bitumen, and plastic polymers are used as matrix materials.

Should recycling and incineration become more widely used in future, it may become feasible to encapsulate the fraction of municipal waste that remains. In this case, the waste would probably need to be sorted and shredded first, like refused-derived fuel.

Containing Wastes

Various materials have been used to encapsulate wastes and render them stable. Below we discuss cement, asphalt, and metals for such applications.

Concrete and Cement

Concrete and cement are the most common encapsulation materials, owing to the availability and cheapness of the raw materials and the durability of the final product. Furthermore, there is a large body of scientific knowledge about cements and concrete stemming from their widespread use in engineering and construction. Concrete is used to encapsulate wastes both at low-level radioactive waste storage and disposal sites (for example, the Swedish disposal site described in chapter 11) and at hazardous waste disposal sites (for example, the Alberta Special Waste System at Swan Hills, Alberta, also described in chapter 11).

Concrete is a mixture of Portland cement (made from limestone), aggregate in the form of sand and/or gravel, air, and water. It is temporarily in a plastic state during which time it can be mixed with or molded around waste. The concrete hardens by a complex series of hydration reactions, the principal one being the hydration of calcium silicates to calcium silicate hydrate and calcium hydroxide. Initially, the calcium silicate hydrate forms a gel, which then develops a rudimentary crystalline lattice that gives strength and rigidity to the concrete. The reaction generates heat, which can induce thermal stresses and cracking in the concrete.

Portland Cement Type 1, sometimes with fine aggregate, is commonly used for the encapsulation of radioactive waste (Taplin & Claridge, 1987). Many additives are available to improve or give special properties to the concrete-waste mixtures. Commonly used additives are air entraining agents, most frequently natural and synthetic soaps, to increase the resistance to frost action; and pozzolans (siliceous substances such as fly ash and pumice that react with lime in the presence of water) to reduce the heat of hydration, increase resistance to sulphates, and prevent calcium hydroxide leaching. Disadvantages of the latter are increased shrinkage with drying and reduced durability.

Other admixtures that are being considered are finely ground silica and fly ash, which increase the strength and decrease the permeability of concrete. The addition of blast furnace slag moderates the hydration reactions to give lower temperatures and, consequently, lower thermal stresses. Because fly ash, usually from coal-fired electrical plants, has been shown to improve the quality of concrete, it would appear feasible to encapsulate ash from waste incineration into concrete on a production scale.

Cement and concrete can deteriorate over time, which can lead to a loss of long-term containment effectiveness. In addition, concrete properties can be affected by shrinkage and microbial action.

Asphalt/Bitumen

Asphalt is a black or dark brown derivative from petroleum evaporation and is also known as bitumen. Like concrete, asphalt can be used for both encapsulation of wastes and for barrier systems in a landfill. Asphalt can also be used as a coating on concrete barriers.

The advantages of asphalt are its low permeability to water, relative abundance and low cost, and resistance to most acids, bases, and organic salts, and some organic chemicals. It is, however, susceptible to attack by organic solvents, particularly hydrocarbons, and would not be suitable for encapsulation or as a barrier for wastes with a significant content of petroleum substances.

Metals

Metals can be used as containers or barriers for waste disposal. Because of their relatively high expense, they are not generally used as an encapsulation material, except for specialized cases of small volumes of very hazardous materials. If groundwater will be present in the landfill, either from the outset or at some future time, a metal should be chosen that corrodes very slowly and is sufficiently thick that it will not fail within the desired containment lifetime. Failure usually occurs as the result of localized corrosive attack, such as pitting, formation of crevices, stress corrosion cracking, and intergranular attack.

The following metals have been evaluated for use as containers or overpacks for containing high-level radioactive waste: stainless steel, inconels (nickel-based alloys), aluminum, monel, copper-nickel, copper, different types of titanium, and carbon steel (Johnson et al., 1994). Carbon steel has received the most attention because it is significantly less costly than the other metals. Stainless steel is of interest because, with its low carbon content, it can be welded without increasing its susceptibility to corrosion.

Although long-term corrosion rates are difficult to predict from short-term testing, results suggest that metal containers, when placed into a well-designed underground environment, may remain unbreached for very long times. In Sweden, it has been proposed to cast high-level nuclear wastes into a lead matrix inside thick-walled (60-mm) copper containers. The containers would be placed deep underground in a borehole and surrounded by compacted bentonite clay. On the basis of current understanding of corrosion behavior in the anticipated groundwater, the time to breach the copper container might approach several tens of millions of years (SKB, 1992).

Vitrification or "Glassification"

The use of vitrification was first explored in the nuclear industry as a means of encapsulating long-lived radionuclides prior to disposal. The thought was that if the radioactive materials are embedded in specially formulated glassy material, they will be contained for very long periods of time, perhaps even approaching the scale of geologic time.

At present, vitrification is not an economical or practical method for treating municipal or hazardous wastes. If new thermal technologies evolve, or if new methods of incineration are developed that operate at very high temperatures, it might become possible to melt ash and incorporate it into a glass or ceramic form. Such waste forms would be extremely resistant to leaching and would be well suited for land disposal.

Discussion Topics and Assignments

1. How long do you feel that the following manmade materials would last as a landfill liner before losing their integrity: 80 mil HDPE, 30-cm-thick reinforced concrete, and 1-m thick compacted clay? Explain your answers.
2. If cost was not a constraint, how would you design a landfill using any of the materials described in this chapter?

Suggested Reading

Cottrell, A. 1988. *Introduction to the Modern Theory of Metals.* London: Institute of Metals.

Double, D. D., and A. Hellawel. 1977. *The Solidification of Concrete. Scientific American*, July, 82–90.

Gillott, J. E. 1987. *Clay in Engineering Geology.* 2d edn. New York: Elsevier.

Fowden, L., et al. (eds.). 1984. *Clay Minerals: Their Structure, Behaviour and Use.* London: Royal Society.

Johnson, L. J., et al. 1994. *The Disposal of Canada's Nuclear Fuel Waste: Engineered Barriers Alternatives.* Atomic Energy of Canada Limited Report, AECL-10718, COG-93-8.

Koerner, R. M. 1994. *Designing with Geosynthetics.* 3rd edn. Englewood Cliffs, N.J.: Prentice Hall.

Neville, A. M. 1981. *Properties of Concrete.* 3rd edn. London: Pitman.

Street, A. C., and W. Alexander. 1972. *Metals in the Service of Man.* New York: Penguin.

11

CASE HISTORIES

Theory is fine, but practical experience is the heart of real learning. This chapter provides—as much as a book can—some real-life experience through seven case histories of how wastes are managed. The case histories describe a state-of-the-art materials recycling facility, five waste disposal facilities in three different countries (the United States, Canada, and Sweden), and a large mass-burn incinerator.

Choosing which of the many thousands of landfills in existence to include was a difficult task. Three municipal solid waste landfills are described. The first, Fresh Kills landfill in New York City, was constructed in 1948 and represents older landfill technology. The second, a new landfill in East Carbon County, Utah, was built in 1992 and incorporates the latest engineered barriers and features of a modern landfill. The third is being developed in a large, abandoned open-pit mine in California. In addition, we discuss a landfill and treatment center for hazardous waste, located in Swan Hills, Alberta. A unique Swedish facility for disposing radioactive wastes rounds out the suite of landfill case histories; this facility takes a very innovative approach to waste disposal and is included to provide a different perspective on this topic.

State-of-the-Art Recycling: The Guelph Wet-Dry Recycling Centre

Materials recovery facilities (MRFs) are the vital heart of modern integrated municipal waste management systems. Without MRFs, recycling on any practical scale would not be possible; it is here that recyclable materials are collected and made ready for sale to secondary markets. One of the most innovative recycle centers in North America has recently been constructed in the city of Guelph in southern Ontario (Guelph, n.d.). It offers good insight into what can be achieved through recycling, and the equipment that is involved.

The city of Guelph, with a population of 95,000, is situated about 60 kilometers west of Toronto. In the mid-1980s, the city began studying ways to reduce the amount of waste being placed in its landfill. These studies received a major impetus in 1991, when the province of Ontario developed a waste reduction plan that required municipalities to reduce the amount of garbage being placed in landfills by 50% by the year 2000. A number of pilot studies were conducted before the present approach was selected.

Guelph's wet-dry approach is different from the blue-box method that is being used in most other North American communities in that it requires residents to divide their waste into two main streams: wet waste (i.e., compostable waste) is placed in semitransparent green plastic bags; dry waste—waste that is not compostable and contains any of the recyclable products—is placed in transparent blue bags or in labeled containers. Although this method has been used in Europe, this was one of the first applications of it in North America. It is a truly integrated approach which offers considerable control, since everything passes through the Wet-Dry Centre.

An advanced approach has also been taken to the collection of garbage. Guelph uses a fleet of ten custom-designed trucks (see Fig. 11.1). Each truck requires only one operator, who drives and picks up both wet and dry garbage, placing them in separate compartments of the truck. About 50 tonnes of dry waste and 40 tonnes of wet waste are collected by each truck per day.

Construction of the recycle center began in 1993 and took approximately 22 months. The dry recycle plant came on line in November 1995, and the composting facility became operational in February 1996. In 1996, the center operated with 40 to 50 staff and processed 70,000 tonnes of waste (30,000 wet and 40,000 dry), working one shift. A maximum capacity of 135,000 tonnes (44,000 wet and 91,000 dry) per year can be processed, working two shifts.

The cost of the Wet-Dry Centre was $27 million (U.S.), which included pilot studies, permits, the truck fleet, and land costs. The annual operating cost is about $1.5 million. It is projected that once

11.1 Custom-designed garbage truck unloading dry residential waste.

full capacity is reached, the annual operating cost will be $3.7 million. This will be offset by revenue from sale of recyclables of about $4.4 million per year, yielding an operating profit of approximately $.7 million per year. Additional economic benefits include the cost saved by avoiding landfill tipping fees and the long-term costs of maintaining closed landfills, as well as the extra jobs created by the recycle center.

The success of this approach is evident: in 1996, the first full year of operation, the center was already achieving 54% diversion from landfill. In other words, Guelph met and exceeded the provincially mandated recycling target three years ahead of schedule. This compares with a recycling rate of about 10% achieved by their previous blue-box program. Furthermore, the approach is inherently very flexible and offers potential for continued evolution and even greater diversion.

A view of the center is shown in Figure 11.2, and a plan of it in Figure 11.3. Situated on 10 hectares (25 acres) of land, the facility has two main components. The first is a large building which houses the material recovery facility. This building receives and processes "dry" waste. The second main component is the organic waste processing facility. Composed of two attached buildings and an adjoining outdoor asphalt pad, it receives and composts "wet" waste.

11.2   Guelph's Wet/Dry Recycling Centre, looking east.

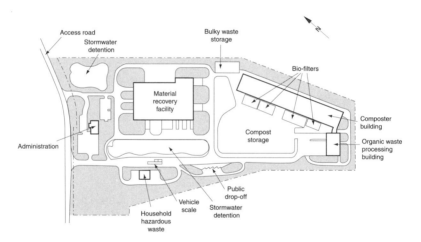

11.3   Plan view of Guelph's Wet/Dry Recycling Centre.

In addition, the center contains a number of ancillary facilities:

- Household hazardous waste facility
- Public drop-off area
- Bulky waste storage area (for large items such as appliances, tires, and furniture)
- Vehicle scale and scale house
- Stormwater management facilities
- Administration building

Siting and Environmental Issues

The recycling center is located on the eastern side of Guelph on land that was purchased from the province. This is a relatively lightly populated rural area, adjacent to an industrially zoned area that includes a provincial correctional center and an abattoir. The site was chosen, in part, to be close to the municipal landfill, which is about 1.5 kilometers away, to minimize transportation of waste. Traffic consists of about 170 to 200 truck visits per day, a relatively small proportion of the traffic on the access road.

To prevent odors and birds—a concern since an airport is located nearby—all waste is unloaded and processed indoors. A baghouse is used to collect dust generated in the MRF. An odor-monitoring program was initially conducted for 9 months at the composting facility and did not reveal any emission problems. No complaints regarding odors have been received from nearby residents.

The surface runoff from the northern half of the site, around the administrative building and the MRF, is collected in lagoons and discharged to the wetlands east of the facility at the rate that would occur naturally if the facility were not present. The surface runoff from the southern half of the site, which includes the composting facilities, is collected in underground wells and recycled by using it to moisten the compost. None of this "gray" water is released to the environment, but excess water not used in the composter is released into the sewer system. Surface water and groundwater monitoring programs have been implemented. The latter consists of regular sampling of six groundwater wells drilled into soils around the site.

Materials Recycling Facility

The materials recovery facility has two main processing lines: one for residential waste, and one for industrial, commercial, and institutional (ICI) dry waste. Pilot studies indicated that ICI waste is much more uniform, consisting mostly of paper and cardboard and with very few containers, bottles, or cans, thus justifying a separate processing line.

The main areas of the MRF are:

- Separate residential and commercial tipping floors
- Manual presorting station
- Mechanical processing area
- Manual sorting stations

- Baling, storage and shipping area for recyclables
- Residual materials (i.e., waste for landfill) storage and shipping area
- Employee service areas (lunch and change rooms)

Residential Waste Line. A flow chart for the residential waste separation process is shown in Figure 11.4, and the floor plan for the materials recovery facility in Figure 11.5, which also illustrates the path that residential dry wastes follow through the facility.

Garbage trucks dump dry waste inside the building onto a concrete tipping floor (Fig. 11.1). A front-end loader pushes the waste onto a conveyor which carries it through a bag breaker to the presort area (Fig. 11.6): cardboard is manually removed at the first presort station, bundles of newspaper at the second, and plastic bags at the third.

The waste is then carried by a vibrating conveyor to a ballistic separator. This machine is set on a slight incline, and the waste passes over a number of revolving eccentric discs. The lighter and flatter "two-dimensional" pieces such as paper and textiles "surf" on top and are carried upslope, and the heavier, rounder "three-dimensional" materials—such as glass, plastic, and cans—go to the bottom and travel downslope to their collection area. Fine materials fall through holes and are collected in a third stream; these are sent to landfill.

The "two-dimensional" waste component is separated manually into the following categories: newspaper, corrugated cardboard, boxboard, fine paper, mixed paper, and textiles. These are dropped down chutes to storage bunkers, from which they are sent to the baler.

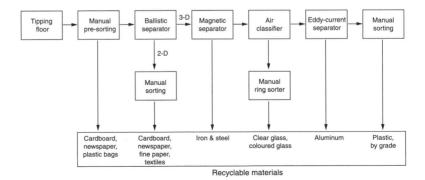

11.4 Flow chart for processing dry residential waste.

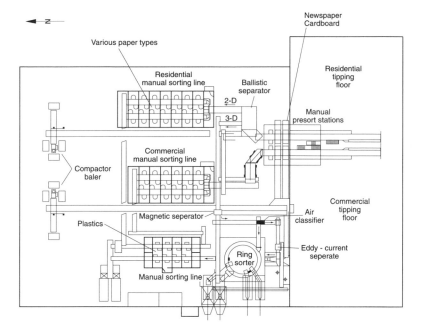

11.5   Floor plan of Guelph's dry material recycling facility (courtesy of City of Guelph).

The "three-dimensional" waste component goes through a sophisticated sorting process:

- An electromagnet separates iron and steel materials
- An air classifier that separates aluminum and plastic from heavier materials, primarily glass.
- Aluminum and plastic go through an eddy-current separator that removes the aluminum. The remaining plastics are then manually sorted into different plastic types.
- Glass materials go onto a ring conveyor where they are manually sorted into clear and colored fractions.

The separated recyclable materials are compacted and baled. Figure 11.7 shows the compactor/baler, and Figure 11.8 shows baled aluminum cans ready for shipment.

Commercial Waste Line.   Commercial dry waste is unloaded onto a concrete tipping floor adjacent to the residential unloading area. The commercial wastes move through a presort station where oversized and problem items are manually removed. The waste then passes over a vibrating screen that removes fine materials, which are sent to landfill.

11.6   Manual sorting of dry waste (courtesy of R. Cave & Associates).

The remaining wastes are sorted manually from a conveyor, with the recyclable materials dropped into bunkers, from which they are sent to a baler. The materials left on the conveyor may go through a second cycle of manual sorting and then pass under an electromagnet which collects ferrous materials. The residual material is compacted and sent to landfill.

Composting

Wet waste is delivered to the receiving building, which has an overall capacity of about 20 to 30 tonnes per hour. Waste is dumped from garbage trucks onto a concrete tipping floor where it is inspected for nonprocessable materials. The waste is then passed to a screw shredder which breaks bags and reduces the size of the pieces.

Then the waste is conveyed into a rotating trommel (Fig. 11.9) which removes large pieces and shredded plastic bags. Oversize

11.7   One of two compactor/balers in the dry material recycling
facility (courtesy of R. Cave & Associates).

materials are passed through the shredder and trommel again; any
remaining oversize materials are compacted and sent to landfill.

Materials passing through the trommel screen are sent under a
magnet to remove ferrous metal. Following this, wood chips or other
additives to enhance the composting process are added, and the wet
garbage is sent to the compost building, which is attached to the
receiving building.

Both the compost and receiving buildings are fully enclosed and
kept at negative pressure to ensure that odors are contained. Air from

11.8   Baled aluminum awaits shipment.

11.9   Wet garbage goes into a rotating trommel that separates oversize materials from the compostable fraction (courtesy of R. Cave & Associates).

11.10 Wet garbage in the primary composting channels. Note the overhead piping for spraying and the compost-turning machine running on tracks set on top of the walls (courtesy of R. Cave & Associates).

the buildings is collected and passed through biofilters consisting of shredded pine bark, chipped hardwood, and leaf compost before being vented to the outdoors.

Composting is done in two stages. The compost building is long and relatively narrow, with eight concrete-walled channels in the front half for the primary composting stage (Fig. 11.10). These channels are sprinkled from overhead pipes with "gray" water to maintain optimum moisture, and aerated from below via pipes in the gravel floor to control temperature and biological activity. The Ontario Ministry of Environment requires that the compost be maintained for at least 3 days at 55°C or higher to destroy pathogens and weed seeds (MOE, 1991). The temperatures in the compost can readily reach higher levels, and it must be cooled. The compost spends 4 weeks in the channels being progressively turned—rolled down the channel—by special machines which run on rails set on top of the concrete channel walls (Fig. 11.10).

The compost is then placed for 4 weeks in two long windrows inside the back part of the building for secondary composting. The compost is aerated from below during this stage and is turned by a front-end loader about once a week. Fresh water is added for moisture.

The finished compost is screened and cured outside on a paved storage area for six months. After this aging, it is sold in bulk for landscaping and site-restoration uses.

Future Directions

In 1996, a total of 16,200 tonnes of recyclable materials were recovered by the MRF, consisting of the following materials (percentages by weight):

- Newspaper (40.7%)
- Corrugated cardboard (25.8%)
- Scrap metal (12.1%)
- Boxboard (5.9%)
- Fine paper (5.4%)
- Ferrous (3.8%)
- Glass containers by color (2.7%)
- Plastic and film (2.5%)
- Aluminum (1.0%)
- Textiles (0.2%)

As shown in Figure 11.11, of the total waste received by the center in 1996, about 43% was wet waste and 57% dry waste. Approximately 67% of the wet waste and 44% of the dry waste were recycled. Of the total waste, about 54% was recycled and 46% was sent for landfill disposal.

The Guelph Wet-Dry Centre is relatively new, and early efforts focused on commissioning the facility and fine-tuning its operations. After three years of operation, the overall waste diversion has stabilized at approximately 58%. A primary objective is to bring the operation up to full capacity by seeking additional waste from the commercial and industrial sectors in Guelph, which are currently sending much of their waste to private landfills, and from the surrounding county, as well as from some institutions outside the immediate jurisdiction.

Guelph's wet-dry approach has enormous potential to decrease the amount of waste being sent to landfill. The system is very flex-

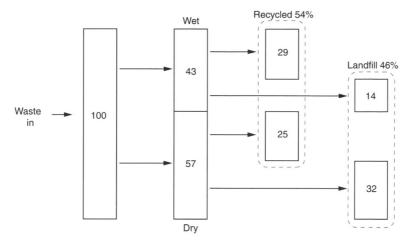

11.11 Schematic diagram showing the recycling efficiency achieved in 1996.

ible, and there is good control because all waste passes through the system. Active research is going on to determine ways of extracting greater quantities of recyclable materials more efficiently.

For example, a good proportion of the waste residue consists of "fines," material of diameter 2 cm or less, that drop through the holes in the ballistic separator in the MRF. Although too small to sort by hand, this material might be well suited to sorting by automated processes which could be implemented either by adapting existing equipment or by adding new equipment.

An exciting development, still in its initial stages, is seeking out niche markets and then conducting selective waste sorting to supply them with the specific raw materials they need. The Guelph facility is well suited for this kind of custom work. For example, industrial plastic film is separated and supplied to a local cottage industry that makes special pillows. ARC Industries Limited, a job provider for handicapped people, is also tapping into the vast potential of Guelph's Wet-Dry Centre by providing labor for the customized waste separation. It is envisaged that Guelph's recycle center will become a mine of raw materials that will attract a host of entrepreneurial cottage industries.

The ultimate step toward achieving the elusive dream of almost complete recycling, however, may lie in a custom-designed anaerobic digester which is to commence operations in late 1999. The digester will produce methane, which will be used to generate electrical power. The digester can decompose plastics, leather, rubber, and other materials currently being sent to landfill. Once this concept comes to fruition, Guelph will achieve over 95% waste diversion—a giant leap forward for the environment.

A Monster: The Fresh Kills Landfill

With a population of more than 17 million, New York City is one of the largest cities in the world and deals with monumental volumes of household garbage. A review of the waste situation in New York City provides a good example of the historical development of waste disposal in North America, as well as a graphic illustration of the inherent problems of large municipal landfills.

An excellent review of the history of waste landfilling in New York is provided by Walsh (1991). Before the turn of the century, New York's solid waste was simply dumped at sea. This proved a cheap and convenient solution until beachfront communities began to protest against the frequent washing ashore of garbage. The solution was found in 1895 in New York's first comprehensive waste management plan. Instead of being towed to sea, barges laden with garbage were taken to landfills located on tidal wetlands skirting the ocean. This was considered a good solution at the time because tidal wetlands were abundant and were thought to be of little value. Filling marshes was believed to improve the economic value of these lands; people did not understand the vital role of tidelands in the ecosystem.

Interestingly, the 1895 waste management plan included an intensive recycling program for paper, wood, metal, cloth products, and food waste. This lasted until around 1918 and was not revived until the current wave of recycling began in the mid-1980s.

After World War I, the volume of waste increased by 70%, with annual waste generation exceeding 11 million cubic meters. To accommodate this waste load, new landfills were opened throughout the city; by 1934, 89 truck-accessed landfills were operating. As the population continued to grow and the geographic limits of the city expanded, it became more and more difficult to site new landfills, and the existing ones came under increasing pressure to operate cleanly or to close altogether.

To that time, the landfills had been only about a meter thick and had grown horizontally, consuming large tracts of land. This growth, along with the inherent dirtiness of landfills, inexorably led to a conflict with urban development. In 1949, the height of landfills in New York was increased to create additional storage capacity. Vertical growth has since become a cornerstone of today's municipal landfills.

Incineration also played a role in garbage management during the 1950s and 1960s. But owing to public opposition, the number of incinerators decreased from 17,000 residential and 22 municipal incinerators to zero residential and three municipal incinerators in 1985 (Neal and Schubel, 1987). The last municipal incinerator was closed in 1993.

The conflict between municipal development and landfills continued to worsen. It combined with vigorous local opposition—the NIMBY syndrome—so that the number of landfills in New York City diminished steadily, from 89 in 1934 to a single one in 1991.

The lone survivor, the Fresh Kills Landfill on the west side of Staten Island, is understandably gargantuan. It now covers approximately 1,200 hectares (3,000 acres) and contains over 70 million cubic meters of refuse. Fresh Kills Landfill has grown vertically to the amazing height of 150 meters. It is difficult to imagine the immense size of the landfill, which is one of the largest structures ever constructed by humans—it is 25 times larger in volume than the Great Pyramid of Giza! In fact, Fresh Kills Landfill is destined to become the highest point of land on the Eastern Seaboard.

The location and site plan of Fresh Kills landfill are shown in Figure 11.12. The landfill has excellent transportation links, with access by barge and boat on the west to the Arthur Kill waterway which connects to the other New York boroughs. An early view of Fresh Kills Landfill is shown in Figure 11.13. In addition, Interstate 440 cuts across the middle of the landfill in a north–south direction. The landfill is divided into four sections or cells, designated 1/9, 2/8, 3/4, and 6/7. Of these, cells 1/9 and 6/7 are still active and continue to receive waste by both barge and truck. Sections 3/4 and 2/8 received waste only by truck and stopped receiving waste in 1992 and 1993, respectively.

Fresh Kills Landfill is owned by New York City and operated by the Department of Sanitation, with a staff of more than 500 people. It services New York's five boroughs with a population of about 7.5 million. Garbage from Staten Island is trucked directly to the landfill, and refuse from the other four New York boroughs arrives by barge. A crushing and screening facility processes about 680 tonnes per day of construction debris, including concrete, asphalt, and soil. This material is used to construct landfill roads and to provide cover

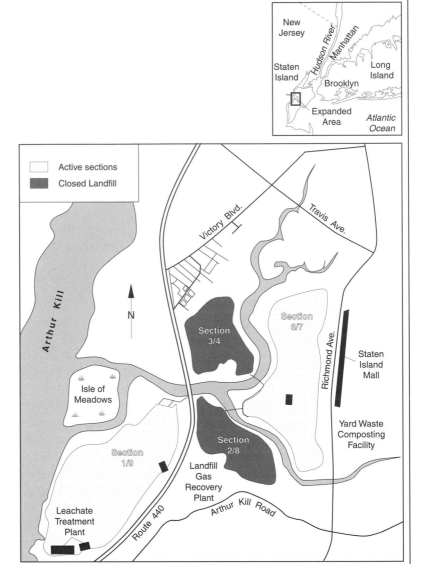

11.12   Location and site plan for Fresh Kills landfill.

material. Landscaping and wooded berms are used to screen the landfill from nearby communities.

Sited on a salt marsh in 1948, the landfill lacks many modern pollution controls, and its management poses complex environmental problems. The Fresh Kills Landfill has no bottom liner to prevent escape of leachate, and because of this, the landfill discharges

11.13 Aerial view of Fresh Kills landfill circa 1943 showing marine unloading plant and barges (courtesy of the Municipal Archives of the City of New York).

at least four million liters of leachate into New York Harbor every day. The tides ebb back and forth at the bottom of the landfill unceasingly, helping to promote biodegradation and leachate generation. A leachate mitigation program was initiated in the early 1990s. A new leachate treatment plant with the capacity to treat 750,000 liters of leachate per day has been operating since January 1994 to help prevent surface and groundwater contamination (New York City Department of Sanitation, 1994a). The plant is designed to remove ammonia, organic matter, and several metals, which are the primary contaminants in the leachate. Work is in progress to expand the leachate treatment capacity to 3.8 million liters per day so that all leachate from the active sections of the landfill will be treated. Leachate from closed sections of the landfill is not being treated, since those sections have been capped and it is anticipated that their leachate generation will decrease.

There are also concerns regarding the stability of the landfill because of its great height. Several hundred borings were drilled through the bottom of the landfill to determine the nature of foundation soils and to collect samples for laboratory testing (New York City Department of Sanitation, 1994b). Numerous analyses of soil and slope stability were performed. In addition, 90 stability-monitoring instruments have been installed which are connected to an automatic data-acquisition system that allows continuous monitoring and assessment of the data. Three types of instruments are used: piezometers measure how fast water is draining from the soils—information necessary to calculate stresses; inclinometers measure changes in the shape and compression of soils and refuse; and temperature sensors measure the heat rise in the landfill, which affects the other measurements, as well as soil behavior. These data provide confirmation that the soils and landfill are behaving as predicted. Seismic stability studies have also been conducted.

It is estimated that the Fresh Kills Landfill emits 1.7 million standard cubic meters per day (MSCMD) of landfill gas, of which slightly more than half is methane. Approximately 2,800 cubic meters per day are odor-causing gases, primarily volatile organic compounds. A major gas management program has been instituted which includes more than 160 continuous methane monitors in on-site buildings, seals installed along underground utility pipes to prevent gas migration, and more than 3.4 kilometers of vent trenches installed around the perimeter of the site to prevent migration of gases off-site (New York City Department of Sanitation, 1994c).

About 0.28 MSCMD (or 17 percent of the total) of landfill gas is collected from approximately 100 wells over 160 hectares (400 acres) of the landfill and processed. The refined pipeline-quality gas (0.14 MSCMD) is sold to the Brooklyn Union Gas Company, a local utility. This initiative, which generates close to $1 million per year in revenue as well as protecting the environment, will be expanded throughout the landfill in the next few years.

Fresh Kills is scheduled to close in the year 2001. At that time state regulations require that it be capped with clay and covered with 0.75 meters of earth. The New York City Department of Sanitation is required to monitor the cap and manage the gas and leachate collection systems for 30 years following closure. Currently, more than 50 hectares (120 acres) have been closed, capped, and covered. The New York Department of Sanitation is working with landscape architects and horticulturists to refurbish the landscape into something both beautiful and useful (Duffy, 1994). Studies are being conducted to identify plant varieties that need little maintenance, will adapt to the landfill's unique conditions, and will encourage use of the area as wildlife habitat.

After recycling, about two-thirds of New York's waste (about 7,700 tonnes per day) goes to Fresh Kills Landfill. One-third (about 3,800 tonnes per day) is being exported to landfills in Virginia and Pennsylvania. None is being incinerated.

This leaves New York City with very few options: NIMBY will not let them site new landfills; there is strong opposition to incinerators; and recycling can divert but a fraction of the waste.

Once Fresh Kills Landfill is closed in 2001, all waste will have to be exported, leaving New York City in a very vulnerable position. It will be reliant on the willingness of other communities to accept its waste. With increasing population and with growing concern about the environmental impact of landfills, it can only be a matter of time before communities close their doors to outside waste. Better solutions are desperately needed.

A Modern Nonhazardous Landfill: East Carbon

A state-of-the-art landfill has been constructed in East Carbon, Utah, by ECDC Environmental, a subsidiary of Laidlaw Environmental Services (ECDC Environmental, undated). This case history was selected because it offers a contrast to the Fresh Kills Landfill. Whereas Fresh Kills is located in a tidal marsh, the East Carbon landfill is in a dry, arid inland site (see Fig. 11.14). Fresh Kills is surrounded by a large urban population; East Carbon is in a sparsely

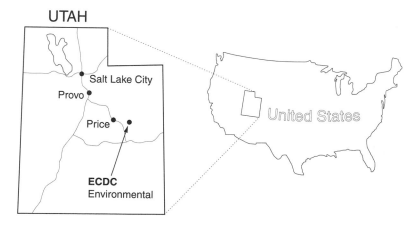

11.14   Location of East Carbon landfill.

populated region. But most significant, Fresh Kills has no bottom liner and has only slowly been incorporating engineered features such as leachate collection and treatment. In contrast, East Carbon has been designed from the outset to include the latest engineered technologies, including a multilayer liner.

However, there is one marked similarity—both landfills deal with enormous volumes of waste. In the case of Fresh Kills Landfill, the large size arose by default because it was impossible to site other landfills in New York City. The ECDC landfill, however, has been designed from the outset to receive enormous volumes of waste.

Completed in 1992, the facility is within the city limits of East Carbon, approximately 225 kilometers southeast of Salt Lake City, in arid high-desert terrain. East Carbon has a population of about 1,400, with coal mining as its primary industry. Initially, there was a vocal group opposed to the town's annexing the land on which the landfill was to be built. A legal suit was launched to stop the process, but it was turned down by the district court. An appeal to the Supreme Court led to the case being returned to district court. In June 1994 a referendum was held in East Carbon; 87% of the voters were in favor of the landfill operation.

The East Carbon landfill has been designed for nonhazardous wastes and can accept municipal solid waste, nonhazardous industrial waste, sewage sludge, asbestos, petroleum-contaminated soils, and ash from incineration of municipal solid waste. At the time this book was written, the landfill was receiving mostly industrial waste.

The East Carbon landfill is unique in that, unlike virtually all other landfills, it is not situated in or adjacent to the urban centers it serves. The location was chosen for its favorable geologic and groundwater features, rather than on the basis of convenience. The landfill is in a relatively sparsely populated area and will rely on waste being transported to it over considerable distance. Thus, this landfill is an example of what may develop in the future: a well-sited large regional facility serving many urban centers.

Rail haulage is a key part of the East Carbon landfill. The site is linked by a 5.6-kilometer rail spur to the nation's railway system, providing economical accessibility for its customers. The landfill is becoming a national resource: it receives wastes from many major centers such as Boston and San Francisco, which are thousands of kilometers distant.

More than 15 kilometers of rail track are installed on the property, and more than 500 railcars and more than 900 containers are owned or leased by ECDC Environmental. The system permits efficient unloading of containers, bottom dumping of railcar hoppers, and rotary dumping of open-top railcars. Figure 11.15 shows railway containers being unloaded onto trucks for delivery to the landfill face.

An area of 960 hectares (2,400 acres) has been permitted for landfill development, with a total capacity of more than 140 million cubic meters. The disposal area is divided into 29 cells of about 32 hectares (80 acres) each. Each cell will be excavated to a depth of 6 meters and the removed soils used to build a 12-meter berm around the cell. Thus, each cell will have a depth of 18 meters and will contain some 4.6 million cubic meters of waste. The landfill will probably have an operating lifetime of 50 or more years. At the time of writing, two disposal cells had been constructed and waste was being received at a rate of almost 2 million tonnes per year. An aerial view of the landfill is shown in Figure 11.16.

The bottom of each cell will be lined with 1.7 meters of natural and engineered materials to form an impervious barrier. This barrier includes about 0.9 meters of compacted clay, two high-density polyethylene liners, and about 0.6 meters of specially screened waste material as protective cover. In addition, the barrier system contains two leachate detection and collection systems to provide

11.15   Containers are unloaded from rail cars for transfer to the landfill face at East Carbon, Utah (courtesy of Brian Kearney, Inc.).

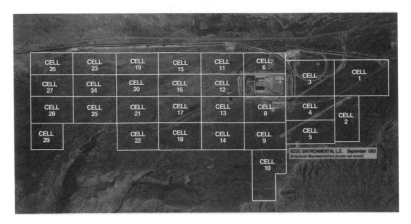

11.16   Aerial view of the ECDC landfill, East Carbon, Utah (courtesy of Brian Kearney, Inc.).

extra safety through redundancy. Their purpose is to detect leaks in the liner system and allow leachate to be removed while preventive actions are taken. (A schematic cross-section of this barrier system is shown in Figure 7.5.) This is a very elaborate liner system; it exceeds the requirements for a nonhazardous landfill and meets or exceeds the requirements for even a hazardous landfill.

After each cell is filled to capacity, it will be capped with a low-density polyethylene liner and an additional 0.6 meters of soil. Native plants will be used to revegetate the cell cover to provide stability and prevent erosion.

The climate and geology at East Carbon provide a good setting for landfill operations. With an average of 28 centimeters of precipitation per year, coupled with an evaporation rate of 120 centimeters per year, there is very little potential for leachate generation from precipitation. Furthermore, the landfill sits on a 450-meter-thick layer of claylike Mancos shale. The water table is at a considerable depth, 1,650 meters below the landfill. However, a perched water table is located about 21 meters below the site. The combination of very deep water tables and low precipitation makes it difficult for water to infiltrate into the landfill and for contaminants to migrate away from the site by groundwater transport, even if they were able to escape from the double liner system. Groundwater monitoring wells have been installed to depths of about 24 to 27 meters in three locations upgradient of the site in terms of groundwater flow, and in three locations on the site surrounding the two disposal cells that have been constructed to date.

No plans have been made for methane extraction, perhaps because of the lack of a nearby market for this energy source.

After the landfill has reached capacity, it will be closed and sealed, and a monitoring program implemented for a 30-year period. Although definite plans will not be made until nearer the closure time, it is currently envisaged that the site will revert to use for cattle grazing.

An Open Pit Megaproject: Eagle Mountain Landfill

This case history illustrates several important new developments in landfill evolution. The first is the emerging use of disused mines for waste disposal. Not only will an otherwise despoiled landscape be reclaimed; in addition, some abandoned mines can contain much larger volumes of garbage than conventional landfills. Second, with the increasing difficulty of siting new landfills, especially in urbanized areas, rail transport is gaining importance for hauling waste to larger landfills in more remote areas. The Eagle Mountain landfill in California's Mojave Desert is a good example of both these trends. It will serve as a regional landfill for seven counties, including the Los Angeles area, to which it is connected by rail. Its vast size will accommodate up to a mind-boggling 635 million tonnes of waste, about five times the size of Fresh Kills Landfill in New York City.

The project is situated in the foothills of the Eagle Mountains near the town of Eagle Mountain in Riverside County, about 320 kilometers east of Los Angeles (see Fig. 11.17). It is jointly owned by Mine Reclamation Corporation and Kaiser Eagle Mountain, Inc., a wholly owned subsidiary of Kaiser Resources, Inc. (Mine Reclamation Corporation, 1995). The latter company mined iron ore from four pits from 1947 until closure in the mid-1980s. The largest pit, one of the biggest in the country, is a vast scar on the landscape: it occupies 320 hectares (800 acres), stretches four kilometers in length and 0.8 kilometer in width, and is 460 meters deep. The pit is shown in Figure 11.18. The landfill includes an additional 520 hectares (1,300 acres) plus a buffer zone of 800 hectares (2,000 acres).

An attraction of the site is the 85-kilometer rail spur that was already in place, linking the mine to the Southern Pacific rail system. In addition, there is existing infrastructure of utilities and roads. A considerable amount of clay and gravel remains from the mining days and can be used in temporary roads and the bottom liner sys-

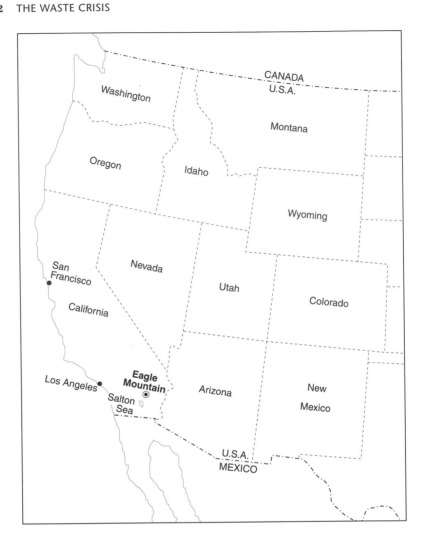

11.17   Location of Eagle Mountain landfill.

tem, as well as for daily cover and final capping, thus minimizing the environmental impact of accessing virgin materials.

The landfill is being permitted as a nonhazardous municipal solid waste landfill and will receive refuse from southern California. No out-of-state waste will be accepted. An important part of the operation is that all wastes must first pass through materials recovery facilities where recyclable and hazardous materials will be removed. The wastes will be transported primarily by rail using sealed and locked containers, although some road transport will be used to

11.18   Aerial view of the former Eagle Mountain mine (courtesy of Mine Reclamation Corporation).

serve the area close to the landfill. The wastes will undergo further inspection once they reach the landfill.

The landfill's footprint will cover approximately 840 hectares (2,100 acres), and the buffer and ancillary areas will occupy another 800 hectares (2,000 acres). Up to 18,000 tonnes of compacted waste will arrive daily. With a total capacity of about 610 million tonnes, the site has an expected lifetime of more than 100 years. Once in operation, Eagle Mountain will be the largest-capacity landfill in the world (Mine Reclamation, 1995).

The landfill is undergoing a lengthy permitting process and is expected to be in operation by late 2000 at the earliest. Six public hearings have already been held, with one more scheduled. Issues raised in the permitting process include potential negative effects on the delicate desert environment, which includes the habitat of the endangered desert tortoise, and on nearby Joshua Tree National Monument. The proponent has been asked to analyze the impact of seismic activity on the landfill, including the effect on the liner system. An interesting commitment that was negotiated with the proponent was that clean-burning natural-gas engines would be used on the trains that haul refuse to the facility.

The geology in the immediate vicinity consists of two major formations. The foothills are composed of ancient granite and metamorphic bedrock. The adjacent valley consists of alluvial sands, silts, and gravels. The landfill is situated primarily in the bedrock, although the eastern portion of the site lies in the alluvial materials. Investigations have shown that there are no recent (within the past 11,000 years) faults located on or near the site.

The site has several features that enhance environmental protection. The annual rainfall in this desert region is a sparse 7.5 centimeters, and much of that evaporates before it has a chance to seep underground. Furthermore, the groundwater table is located 15 to 82 meters below the landfill bottom, depending on location. The groundwater, which is of the calcium-magnesium sulphate type, is contained in fractures in the bedrock and flows to the south and southeast.

To further isolate the wastes from the groundwater system, 2.4 meters of liner is being placed on the mine floor, consisting of clay and two layers of 2.0-mm (80-mil) high-density polyethylene. The system incorporates a collection system so that any leachate accumulating at the bottom of the landfill can be removed immediately via 10 sumps placed at low points in the landfill. Underlying the liner is an unsaturated leachate monitoring system which is designed to detect and collect any liquids that migrate through the liner. It consists of a collection zone, a 2.0-mm (80-mil) HDPE liner, and a clay liner. EPA test models indicate that with the low rainfall in the region, leachate will probably not form, much less leak from the landfill. The liner system exceeds all state and federal requirements and is considered to be unprecedented for a municipal nonhazardous solid waste landfill in California (Mine Reclamation, 1995).

The landfill is designed to handle the large volume of landfill gas that will be generated as the garbage biodegrades. The gas will be collected by a series of horizontal pipes and vertical wells built into the waste which carry the gas to a flare, where it will be burned at temperatures exceeding 780°C. At this point, there is no plan to use the methane to generate electricity, although this decision will be reconsidered at a later date.

About a thousand hectares of unused land will be returned to the Bureau of Land Management as the site is progressively completed over the coming decades. The mining scar will be restored to its natural state by adding native flora to the final cover, which will consist of about 0.6 meters of compacted soil, a very-low-density polyethylene layer, 0.3 meters of granular soil on top of a geotextile cushion, and an erosion-prevention layer of cobbles and boulders. Comprehensive monitoring—consisting of perimeter wells, leak detection below the landfill, and air sampling on all four sides—will continue for at least 30 years after closure.

Eagle Mountain is one of three new waste disposal megaprojects planned for southern California. In the second project, Waste Management of North America, Inc., will ship waste by rail in double-stacked modular containers to a landfill in the Mojave Desert near the town of Amboy. The site occupies 11.7 square kilometers, and its capacity is similar to that of Eagle Mountain. In the third, Western Waste Industries, Inc., plans to place municipal waste into disused parts of a gold mine in Imperial County. The facility will contain up to 365 million tonnes of refuse and will have a lifetime of about 40 years. Waste haulage by rail will be a key part of the operation.

Landfills are essential for managing California's waste because there is fierce public opposition to incinerators. In the past, southern California has relied on numerous landfills near the urban centers they serve. The trend now is toward a few remote mega-landfills, which are regional in scope and rely on rail haulage to deal with the large volumes of waste and great distances involved. This is a trend that will probably be seen throughout North America.

An Integrated Hazardous Waste Facility: Swan Hills

Let us now turn to a hazardous waste disposal facility. In the United States, hazardous wastes are concentrated, put into drums, and buried either in one of 21 specially designed commercial hazardous waste landfills, or in one of 35 similar landfills operated by companies for their own hazardous waste (Miller, 1997). In Canada, three major landfills are licensed for hazardous wastes: the Swan

Hills Treatment Centre near Swan Hills, Alberta; the Laidlaw Environmental Services facility near Sarnia, Ontario; and the Stablex Canada, Inc., facility in Blainville, Quebec. The first two facilities also incinerate waste.

The Swan Hills Treatment Centre, situated in Swan Hills, Alberta, about 210 kilometers northeast of Edmonton (see Figure 11.19), is the most comprehensive hazardous waste treatment and disposal facility in North America. It is a fully integrated operation which can incinerate organic liquids and solids, treat inorganic liquids and solids, and provide landfill disposal of contaminated solids or deep-well injection disposal of contaminated liquids. The facility has been one of the few to have overcome the NIMBY syndrome in recent times, and it serves as a case study in how to site waste management facilities. (This subject is addressed further in chapter 12.) The facility is owned and operated by Chem-Security (Alberta), Ltd. Originally ownership was in a joint venture arrangement with the province of Alberta, but the province recently sold its share of the venture to Chem-Security.

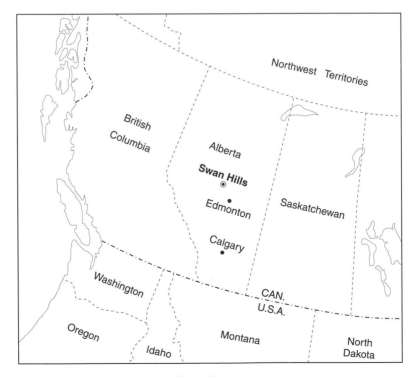

11.19   Location of Swan Hills, Alberta.

Opened in 1987, the facility handles all types of hazardous wastes except pathogenic, explosive, or radioactive materials. It can process more than 100,000 tonnes of waste annually. Advanced receiving, storage, and feed preparation systems facilitate the efficient handling and processing of liquids, solids, and sludges. A view of the facility is shown in Figure 11.20. Originally intended solely for provincial use, Swan Hills opened its doors to hazardous wastes from the rest of Canada in February 1995.

The Swan Hills facility is the central component of a comprehensive hazardous waste management system for the province, which includes a transportation network and waste transfer and storage facilities in two other locations in Alberta, as well as an extensive program that encourages waste recycling and on-site waste reduction (Rabe, 1994b).

Incineration and Waste Treatment

Organic wastes such as oil, solvents, and PCBs are treated by high-temperature incineration to achieve 99.9999% destruction of hazardous materials. Two rotary kilns are available. The larger one has

11.20   Aerial view of Alberta Special Waste Management facility (courtesy of Chem-Security (Alberta), Ltd).

a capacity of 35,000 tonnes per year and operates in either an ashing or a slagging mode; the latter renders the residue into a solidified matrix. The smaller incinerator, which allows some waste streams such as soils to be treated more cost-effectively, has a capacity of 8,000 tonnes per year and operates only in the ash mode. The equipment includes a spray drier, activated carbon injection, baghouse, saturator, and high-energy scrubber. Combustion gases pass through a high-performance flue gas cleaning system which removes acid gases and solid particulates. The ash from the incineration process is rendered inert, undergoes leachate testing, and is landfilled in secure cells or disposed of by deep-well injection (Chem-Security, undated).

Inorganic wastes such as acids, alkalis, and heavy-metal-bearing materials are treated by various physical/chemical processes: neutralization, the addition of acids or bases to the waste stream to create neutral salt solutions; chemical oxidation/reduction, in which appropriate oxidizing or reducing agents are added to the waste stream to break chemical bonds and convert hazardous components into simpler, less toxic substances; and precipitation, the use of chemical flocculation techniques to separate solids from the liquid component of the solution. The materials resulting from these processes may be sent to a filter press which further separates solids from liquids.

The solid residue, or filter cake, is transferred to a stabilization and solidification plant for analysis and stabilization with appropriate binders and additives. Stabilized wastes are cured for 4 to 28 days, then undergo a leachate test; if they pass, they are placed into landfill.

Given the complexity of the operation, it is understandable that it has suffered occasional operating problems. In 1997, elevated concentrations of PCBs were detected by the environmental monitoring program in off-site air and vegetation samples. The sources were found to be outdoor transformer processing, mechanical failure of a transformer furnace, and fugitive emissions from a tank farm and storage buildings. These processes were immediately shut down and preventive measures put in place (Turner, 1998).

Deep-Well Disposal

Liquid wastes from the treatment facilities and leachate from disposal cells are treated to reach levels prescribed by regulatory agencies and then disposed by injection into one of two on-site wells approximately 1,800 meters deep. The waste fluids are injected into a brine aquifer that is well below the level of any usable water-bearing horizons. Deep-well injection is suited to Alberta because of the expertise and equipment available in the oil-exploration industry, where drilling fluids and associated wastes are commonly disposed of in deep brine aquifers. Deep-well injection is not widely used in the rest of Canada; for example, it was originally used at the Laidlaw hazardous waste facility near Sarnia, Ontario, and by chemical industries in that area, but the practice was discontinued in the late 1970s.

Landfill Disposal

The cornerstone of the Swan Hills Treatment Centre is the secure and final disposal of treated wastes into landfill cells. These cells are built into a thick clay layer which underlies the site. The facility is licensed as a secure landfill for hazardous wastes. The bottom of each cell is lined with a multilayer containment barrier consisting of two high-density polyethylene (HDPE) liners as well as protective geotextile and geogrid layers. Above this liner is a leachate collection system consisting of a sump and piping system so that liquids can be extracted, should any accumulate.

As an additional safeguard, a leachate monitoring system is placed between the two polyethylene layers to check for leaks. Both the leachate-extraction and leak-detection systems are constantly monitored. In addition, an engineered groundwater collection system is installed below each cell to minimize external forces on the cell liners.

During operation, the cell is enclosed under a tentlike temporary structure (see Fig. 11.21) which prevents precipitation from entering the cell and facilitates operation during the cold winter months. The temporary cover is removed once the cell has been closed and capped.

When filled to capacity, a cell is covered with a HDPE membrane which is thermally bonded to the inner cell liner to form a complete watertight membrane around the wastes. In addition, the cell is covered with one meter of compacted low-permeability clay. The clay cap is subsequently dressed with topsoil and vegetated to prevent erosion and improve its esthetic appearance. Closed cells are monitored through a system of external groundwater monitoring wells as well as by the leak-detection system and underlying groundwater collection pipes (Chem-Security, undated).

11.21   Construction of a secure landfill nears completion under a tent at Swan Hills Treatment Centre (courtesy of Chem-Security (Alberta) Ltd).

Although the Swan Hills hazardous waste landfill and the East Carbon municipal waste landfill deal with different categories of waste, there are several similarities in their design and operation. Both facilities are situated in relatively sparsely populated areas and are designed to serve wide geographic regions. The two landfills are similar in design, consisting of cells with elaborate multi-component covers and liners to prevent leachate escape. The East Carbon landfill will contain about 50 times more waste on completion than the Swan Hills landfill.

The major difference between the two facilities is the capability of Swan Hills to treat wastes prior to disposal. The East Carbon facility, like virtually all municipal waste landfills, essentially provides only a disposal service; wastes are emplaced in the form they are received, after sorting for recyclables. In contrast, Swan Hills also provides incineration and sophisticated waste treatment processes, as well as waste stabilization prior to disposal. In other words, the facilities at a hazardous waste disposal site are elaborate and sophisticated so that the hazard of the wastes can be eliminated or reduced to the greatest extent possible. Disposal is used only if no other treatment alternatives are available, and the wastes are encapsulated into a stable cementitious form prior to disposal. In fact, the landfill can be viewed as ancillary to the primary activity of waste treatment.

The Inorganic Waste Dilemma

Because treatment processes—primarily incineration—destroy organics, the majority of waste that is placed into landfill at a hazardous waste facility is inorganic. As discussed in chapter 6, this waste will not decay to innocuous forms over time: it will remain hazardous forever. Thus, hazardous waste landfills should be designed and constructed to provide containment in perpetuity. This has important implications.

All of North America's major hazardous waste landfills are surface facilities that are exposed to ongoing weathering, erosion, and freeze/thaw cycles, which will unrelentingly erode the surface covers of these facilities. Thus, perpetual monitoring and maintenance of the covers will be required.

In addition, synthetic materials like polyethylene liners will eventually degrade and deteriorate. Because such materials are unlikely to provide protection for more than about 50 to 100 years, it is inevitable that at some future time these barrier systems will fail and the liners will leak. Other engineered features such as drainage and leachate collection systems also have limited lifetimes.

The design and regulation of hazardous waste landfills is generally focused on a period of a few decades to a century or so after closure. Planners do not recognize that maintenance is required in perpetuity. This places a significant responsibility and burden on future generations and is not consistent with sustainable development.

Nuclear Waste Disposal: The High-Tech Approach

The next case history describes a revolutionary departure from the classic landfill approach that was developed by Sweden for its low-level radioactive wastes.

Nuclear wastes have received more intensive attention than other waste types, and therefore, approaches to their disposal are more advanced. The reasons are threefold. First, the nuclear industry is a high-technology sector and has tackled the problem with a high-technology philosophy. Second, the quantity of waste is very small compared to the economic benefit gained, allowing considerable financial resources to be applied to the problem. Third, public opinion—which sees all nuclear matters, including waste management, as horrifying and hazardous—has applied enormous pressure on the nuclear industry to be extremely thorough. Public opinion has also influenced nuclear regulatory agencies to make the requirements for radioactive wastes more rigorous than for other wastes.

The Swedish "landfill" is called the Swedish Final Repository and is situated on the country's east coast at the Forsmark Nuclear Power Station, about 160 kilometers north of Stockholm (see Fig. 11.22). What makes the disposal method unique is, first, that it is underground, and second, that it is situated under the Baltic Sea (Carlsson, 1990).

Access to the repository is via two tunnels with their entrances on land. They slope gently downward for a distance of 1 kilometer underneath the Baltic Sea. One of the tunnels is dedicated to waste transportation and is equipped with remotely controlled vehicles.

The repository consists of a large vertical silo (60 meters high and 30 meters in diameter) and four parallel, horizontal caverns which are 160 meters long, 14 to 20 meters wide, and up to 18 meters high. The top of the repository is about 50 meters below the seabed; the depth of the Baltic Sea in this area is about 5 meters. A schematic view of the facility is shown in Figure 11.23.

Construction began in 1983, and the repository started to accept wastes in April 1988. A second phase is planned to add another silo and one or two horizontal caverns in about the year 2000. The facility will take *all* the low-level and intermediate-level radioactive waste generated in Sweden until approximately the year 2010 (SKB, n.d.). The total facility is designed to contain approximately 90,000 cubic meters of waste. This is less than 5% of the capacity of a single medium-sized municipal landfill (2 million cubic meters), illustrating the small volume of nuclear waste.

The placement of the repository under the sea is an ingenious idea and has several advantages over the standard landfill. First, only a minimal amount of surface land is required for administrative buildings, and this becomes available for other beneficial uses once the repository is closed. In addition, the wastes are effectively entombed beneath 50 meters of solid rock. This prevents erosion and other forces of nature from slowly degrading the integrity of the facility, so ongoing maintenance will not be necessary after the repository has been closed. The thick layers of rock and sea also form

11.22   Location of Swedish underground low-level radioactive waste repository.

an effective deterrent against future human intrusion, either intentional or accidental. Finally, in the event that there was some leakage from the repository, it would enter the seawater and would not impair any drinking-water resources.

A box within a box within a box within a box, the Swedish Final Repository was designed to prevent radioactive materials from escaping to the environment in harmful quantities, even long after the facility has been closed. This has been achieved by surrounding the wastes with a number of barriers, both natural and engineered.

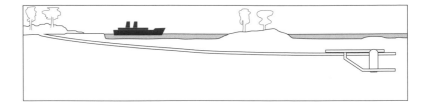

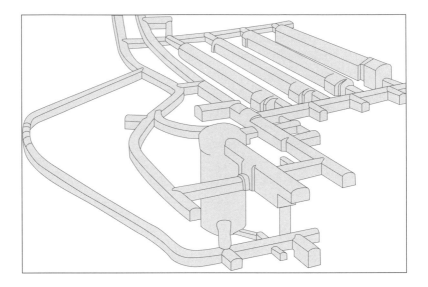

11.23   Schematic view of the Swedish underground low-level radioactive waste repository.

The first barrier is the waste form itself. Before transportation to the repository, the waste is mixed with cement or asphalt and placed in concrete or metal containers. Other wastes that contain some entrained liquids are placed in concrete tanks and dewatered. Less radioactive wastes are compacted into bales or metal drums and placed in steel freight containers. Incinerator ash is immobilized into concrete in metal drums. In comparison, nonradioactive municipal or industrial wastes rarely receive immobilization or pretreatment prior to disposal.

The silo, which has 0.8-meter-thick reinforced-concrete walls, contains the most highly radioactive materials, while materials of lower radioactivity are placed in the tunnels. The space between the concrete silo and the rock wall is filled with bentonite clay, which expands when it becomes wet and will provide a protective barrier against the escape of radioactive contaminants. The concrete silo provides another barrier. The inside of the silo has been divided into square vertical pits into which waste containers are lowered from the top. As each layer of waste is emplaced, it is grouted permanently into place with concrete; this grout provides another barrier. Thus, the engineered barriers consist of the immobilized waste form, grout, the silo wall, and the bentonite layer. Similar barrier systems are used in the four horizontal caverns, with some variations depending on the level of radioactivity of the wastes.

The surrounding rock mass provides an additional important barrier. It consists of the very hard, insoluble rocks of the Baltic Shield, which have been shown through an intensive exploratory program to have very few cracks and fissures and to have low hydraulic conductivity. This kind of rock contains only small amounts of groundwater which travels slowly. The "box within a box within a box" concept is shown schematically in Figure 11.24.

The wastes inside the Swedish Final Repository will decay to about the same level of radioactivity as the surrounding rocks in about 500 years. Less than 10% of the radioactivity will remain after 100 years. This is of the same order of time that it takes for the organic compounds to decompose in a municipal or industrial landfill. It is interesting that the hazardous lifetimes of municipal/industrial landfills and the Swedish repository are comparable, yet the approach to containing the wastes over this timespan are quite different.

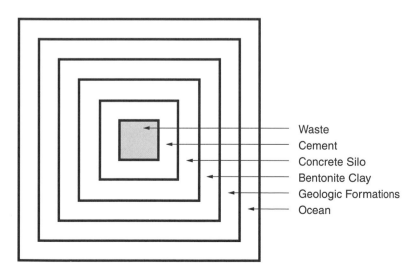

11.24   Waste containment in a box within a box within a box.

It should be noted that this facility is only for low-level and intermediate-level radioactive wastes. High-level nuclear wastes, which contain much higher levels of radioactivity and have a much longer hazardous lifetime, will be disposed of in similar underground caverns but at a greater depth of about 500 meters.

The Swedish approach differs dramatically from what has historically been done in waste disposal. Why did they choose this particular concept? What lessons can be learned?

The two approaches are compared in Table 11.1. It is important to realize that the hazards contained in the two types of disposal facility are fairly similar. As we saw in chapter 6, the contents of municipal landfills are by no means benign. The concentration of hazardous substances in municipal and industrial landfills is considerably lower than in radioactive wastes; however, the total volumes of municipal wastes are much higher. When concentration is multiplied by volume to obtain the total hazard, the two types of disposal facility do not differ greatly. It is interesting to note that, although the biological effects of the two types of waste are similar, the public views nuclear waste with considerably more fear.

Radioactive wastes will be contained in multiple barriers that make the contamination of groundwater very unlikely, whereas surface landfills have a high potential for leakage at some point during their hazardous lifetime. The Swedish repository utilizes

Table 11.1.   Comparison of a typical municipal/industrial landfill with the Swedish Final Repository.

	Typical Landfill	Swedish Repository
Hazard level	Medium	Medium
Hazard decreases with time	Yes, organic decomposition	Yes, radioactive decay
Hazardous lifetime	Organics: 100 to 200 years Inorganics: forever	200 to 500 years
Capacity	Up to a few million m^3	90,000 m^3
Number of facilities	Thousands	One for all Sweden
Waste containment	Cover and liner	Multiple barriers: concrete, bentonite, 50 m of rock, ocean
Ongoing maintenance	Yes	No
Other benefits	Energy from methane	None
Potential to pollute air/groundwater	High	Very low

only minimal land surface and places no burden on future generations. In a nutshell, the Swedish approach is inherently safer and
more attuned to the principles of sustainable development than its
counterpart, the surface landfill. It is a bit of an enigma why the two
waste types are dealt with so differently.

How has this happened? There appear to be two reasons. First,
the Swedish program started from scratch with the specific objective of devising a safe disposal system. In contrast, the landfill has
evolved slowly, with its earliest primary motivation being convenience rather than safety. The modern landfill carries with it
genetic, inherited flaws. Another factor is the large disparity between the volumes of the two kinds of waste. The very small quantities of radioactive wastes have made it possible to spend large
sums of money on research and planning to develop the undersea
concept.

Two important lessons emerge from this case history. First, disposal of wastes deep underground has a number of significant benefits. Second, if waste volumes are relatively small, more sophisticated disposal alternatives become feasible. This emphasizes the
importance of reduction, recycling, reuse, and treatment programs
to decrease the volumes of municipal and hazardous wastes that
must go to disposal. Such programs are vital if safe disposal methods
are to be developed.

Burn Baby Burn: The Lancaster County Incinerator

Tucked into the rolling agricultural countryside of southern Pennsylvania is a large, modern building with clean lines and pleasantly
landscaped grounds. Only its 93-meter stack gives a clue that this
is the Lancaster County municipal waste incinerator. Figure 11.25
shows the location, and Figure 11.26 is an aerial view of the facility.

In 1986, the rapidly increasing amount of refuse led the Lancaster
County Solid Waste Management Authority to adopt an integrated
waste management system using the latest technologies and methods.
With a population of more than 420,000 in a largely agricultural
region, a prime goal was to save valuable farmland by decreasing
reliance on landfills.

The integrated system is based on three main components: re-
cycling/reduction, incineration, and landfilling. Landfilling was
designated the choice of last resort, with the intent of using only
one landfill in the county and making it last as long as possible.

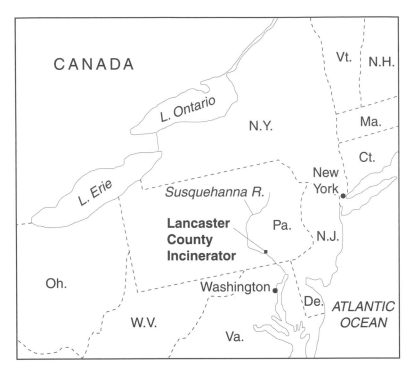

11.25 Location of Lancaster County waste incinerator.

11.26 Aerial view of Lancaster County waste incinerator (courtesy of Lancaster County Solid Waste Management Authority).

The authority felt that its objectives could not be met by recycling alone, so an important component of the strategy was incineration. Ogden Martin, Inc., a major international supplier of municipal mass-burn incinerators, was selected to provide the plant. The Lancaster County incinerator (Lancaster County Solid Waste Management Authority, n.d.; Flosdorf & Alexieff, 1993) is one of 28 Ogden Martin incinerators in the United States, with several hundred elsewhere in the world.

Licensing of the Lancaster County incinerator took 2 years. The final site in Conoy Township was one of 15 sites that were originally considered. A license application, including an environmental impact statement, was submitted to the regulatory agency, the Pennsylvania Department of Environmental Resources; this detailed the operation of the facility, its air and water emissions, and other impacts on the local environment, such as traffic and noise. As happens with the proposed construction of any major facility today, opposition was raised by some local citizens and activist groups. Most of the complaints involved the potential impact of increased truck traffic. A traffic study was undertaken that included assessing the feasibility of using rail transport. Disposal of ash was also an issue with the public.

Groundbreaking took place in March 1989, and just over 2 years later, in May 1991, commercial operation of the plant commenced. The incinerator was designed and built by Ogden Martin Systems of Lancaster, Inc., which also operates the plant; it is owned by the authority. In 1997, six years into the plant's 35-year design lifetime, the authority was very pleased with its performance.

The incinerator is a mass-burn type, based on German technology and consisting of three identical boilers. (A schematic cross-section of the incinerator is shown in Figure 9.1.) Waste arriving at the site is weighed at the scale house and the information is entered into a computer system. The waste is then dumped onto the tipping floor, where it is inspected by trained compliance officers before being pushed into a large pit with a capacity of 3,960 tonnes. Overhead cranes use giant grapples to mix waste in the pit to dry and homogenize it. The waste then goes onto grates in the furnace. The waste burns for about 45 minutes while traveling down the inclined grates, which also agitate the waste. Propane is used to start the incinerator and also (infrequently) when the waste is wet and supplemental heat is required to reach the necessary temperature. Air is blown under and around the waste to ensure complete combustion. It is a regulatory requirement that combustion must proceed at a minimum of 1,000°C. The ash that reaches the bottom of the grate drops into a trough where it is quenched with water.

Managing the gases produced by combustion is an important feature of the incinerator. By the time the gases leave the boiler, their temperature has dropped to about 235°C. The gases then pass into a dry scrubber (one for each furnace), where they are injected with a lime slurry to neutralize acid gases such as sulphur oxides and hydrogen chloride. Next, the gases enter a fabric filter baghouse (one for each furnace) for removal of particulate matter. The scrubbers and baghouses are shown in Figure 11.27. The remaining gases then are discharged through the 93-meter-tall stack.

The plant is operated from a control room equipped with four independent computer systems, one for each furnace plus one back-up. The operators can view different parts of the facility by using remote video cameras.

Ash from the dry scrubber and the baghouse (fly ash) is mixed with hydrated lime and water to prevent escaping dust and is then combined with the ash from the grates (bottom ash). This is sent to a residue building where a rotating-drum magnet removes ferrous metal for recycling. The ash, which represents 10% by volume of the original waste, is sent for disposal at Lancaster County's landfill.

11.27   Rear of incinerator with scrubbers, baghouses, and silo for storing lime.

Steel tubes containing water form the inside walls of the so-called "waterwall furnace." Water inside these tubes is converted to steam, which is further heated in a superheater before it drives turbine-generators to create up to 35.7 megawatts of electricity (4 to 5 megawatts are used to power the plant).

The plant, which has a capacity to burn 1,100 tonnes of refuse per day, operates 24 hours a day. Major maintenance is performed during spring and fall outages that last about 7 to 10 days. The incinerator generally averages less than 10% downtime.

Pollution Control

The facility consumes 1.9 million liters of water per day for steam production, cooling, ash quenching, and other uses. A nearby sewage treatment plant supplies the water, which was previously discharged to the Susquehanna River. The water on site is treated and recycled—no wastewater leaves the site. Two ponds with a total capacity of 9.5 million liters store treated water.

Effective and complete combustion is the most direct way to prevent airborne contaminants. This is achieved in the furnaces by

- Supplying adequate air—that is, oxygen
- Maintaining temperatures above 1,000°C
- Ensuring that waste is burned for a sufficient time
- Mixing the waste during combustion

When proper combustion conditions are maintained, over 99% of the combustion gas leaving the boilers is composed of oxygen, nitrogen, carbon dioxide, and water. The remaining 1% of the gas consists of various pollutants. Pollution-control equipment includes dry scrubbers and baghouses. A continuous emission-monitoring system samples and analyzes stack emissions and automatically adjusts the pollution-control equipment. Combustion efficiency, carbon monoxide, hydrogen chloride, nitrogen oxides, opacity, and temperature are monitored. The Pennsylvania Department of Environmental Resources has ongoing, real-time computer access to these data via a modem link.

The operating emission parameters that must not be exceeded are:

- Sulfur dioxide: 30 ppm hourly; 29 ppm 3-hourly averages; 90% reduction on an 8-hour average
- Carbon monoxide: 200 ppm hourly and 200 ppm 8-hour average

- Nitrogen oxides: 300 ppm daily average
- Opacity: 10% for more than 3 minutes in any hour, never to exceed 30%
- Hydrochloric acid: 30 ppm hourly average, or 90% control hourly—i.e. removal between economizer and stack
- Dioxins and furans: (expressed as 2,3,7,8 TCDD equivalents) 2.0 nanograms per normative cubic meter, with an annual ambient concentration of 0.03 picograms per cubic meter as predicted by computer modeling
- Particulate matter (particles less than 10 microns in size, or PM_{10}): 0.353 grams per cubic meter
- Particulate matter (total suspended particulates, or TSP): 0.424 grams per cubic meter

Costs

Economic viability is a key issue, since an incinerator involves complex technology and large up-front costs compared to a landfill. The capital cost of the Lancaster County incinerator, including design, licensing, and construction, was $105 million. The operating cost is about $9.5 million per year.

User fees finance all of the authority's activities, including the incinerator. In addition, revenue is earned from the electricity generated, which is purchased by Metropolitan Edison for about 5.7 cents per kilowatt hour. The approximately $12 million (about $17 per tonne of waste) that is earned each year is shared between the authority (90%) and Ogden Martin (10%). Ferrous metal recovery earns about $200,000 a year. Additional revenue is earned by incinerating waste from outside the county.

In summary, the authority feels that the overall impact of the incinerator on the management of waste in Lancaster County is quite positive. The cost of waste incineration, when electricity is generated and sold, is competitive. Furthermore, waste incineration to date has worked efficiently, with only about 10% downtime.

Conclusion

A very important outcome is that the incinerator has *not* impeded the county's recycling programs, contrary to the predictions of groups that oppose incinerators. Currently, Lancaster County is recycling about 31% of its municipal waste, an increase of about 10% over the past five years. In fact, incineration has improved overall recycling by recovering ferrous metal that would not otherwise have been reclaimed by the blue-box program. It is clear that incineration and recycling, when designed and operated as part of an integrated system, are complementary and form a powerful team in reducing the amount of waste that is sent to landfill.

Landfill space has also been saved by "mining" and incinerating wastes from the landfill, as discussed in chapter 8. Not only was a complete cell of 7.2 hectares (18 acres) reclaimed; this activity also allowed engineers to inspect and improve the liner of the cell, enhancing the protection of the environment.

In 1996, Lancaster County recycled 31% of its waste, incinerated 56%, and landfilled 13%. This is a very impressive achievement.

Discussion Topics and Assignments

1. If you were director of New York City's waste management program, how would you deal with waste after Fresh Kills Landfill closes in 2001? Explain your ideas.
2. Of the seven case histories, select the facility closest to you, and contact its management to learn how it has evolved since the writing of this book. If it is practical, visit the facility and take a tour.
3. Create your own case history by selecting a nearby waste management facility of interest. Compile information by visiting it, obtaining brochures and reports, and interviewing staff.

12

THE ALL-POWERFUL NIMBY

Our society has reached a frustrating impasse: everyone wants consumer goods, but nobody wants the associated waste. In all levels of society from the grass-roots to the highest level of politics, enormous public opposition has developed to siting landfills, incinerators, or transfer stations. With complex judicial and political systems that promote empowerment of the people, it has become common for opposition groups to delay or halt altogether the introduction of new waste management facilities. The NIMBY—Not In My Back Yard—syndrome has become a powerful force.

Acronyms Express a State of Mind

NIMBY = Not In My Back Yard
NIMTOO = Not In My Term Of Office
LULU = Locally Unacceptable Land Use
BANANA = Build Absolutely Nothing Anywhere Near Anyone
NOPE = Not On Planet Earth

This chapter explores the process by which the sites for landfills and related waste facilities are selected. This fascinating topic goes far beyond technical issues: it provides insight into human behavior and the ways political decisions are made. An understanding of the NIMBY phenomenon is essential for anyone who wishes to pursue a career in waste management.

In some regions there is already a crisis. In New Jersey, for example, the number of landfills has dropped from more than 300 to about a dozen in the past two decades. As a result, more than half of New Jersey's municipal solid waste must be exported to other states. In New York state, 298 landfills were closed and only 6 new ones opened in the decade since 1982. The same story is unfolding in almost all jurisdictions in North America; the number of landfills in the United States dwindled from 20,000 in 1979 to about 5,300 in 1993 (Miller, 1997). There is a very strong trend toward fewer—but much bigger—landfills. In the United States it is estimated that 8% of the existing landfills handle 75% of the country's garbage. As the number of landfills decreases, their heights grow, casting dark shadows across the land.

There is no doubt that new landfills are safer than old ones: they are generally better sited and incorporate better engineering and modern technology such as liners, covers, and leachate and gas extraction systems. However, people still do not want them next door. Thus, the few new landfills that are being developed are getting larger and larger; the megadump is the trend of the future.

The difficulty of siting waste facilities has had a number of negative effects. A direct economic spinoff has been an increase in tipping fees. This is simply an expression of the law of supply and demand: as landfill space scarcer, its cost increases. This in turn has resulted in a proliferation of illicit waste dumping, the so-called "midnight tipping."

Environmental Scams

Illegal or "midnight" dumping proliferates whenever landfill tipping prices rise. In the mid-1990s, numerous illegal dumps were established and operated without permits. Regulatory agencies reported that padlocks on gates were being broken to gain entry to vacant lands for dumping waste. Several cases have been reported in which warehouses were rented, stuffed full of waste, and then abandoned.

The Causes of NIMBY

The siting process has evolved substantially in the past decade. As the costs, energy, and frustration of trying to site unwanted facilities have increased, more effort has gone into trying to understand the NIMBY syndrome. Considerable intellectual effort has been invested in studying several problems: how to increase public involvement and transparency in the siting process; how to create a process that builds consensus instead of confrontation; how to disseminate information effectively; and how to design fair compensation programs. As a result, the siting of controversial facilities has undergone considerable change. For example, it is now routine to offer host communities comprehensive compensation packages that include not only economic benefits but also provisions for health and safety protection and a role in decision-making regarding the proposed facility. This is quite a turnaround from the days of "decide-announce-defend."

Yet in spite of these advances, progress has been minimal. Let us explore why the NIMBY syndrome takes such a powerful grip on the citizens of involved communities.

First, it must be recognized that landfills and other waste management facilities have a notorious history of leaking and causing environmental problems. For example, one-fifth of all Superfund hazardous waste sites are former municipal landfills (Miller, 1997). Even though technologies have improved, memories of the past linger.

Furthermore, even with the best pollution-prevention measures, landfills have other negative aspects, such as truck traffic, flocks of seagulls, or windblown litter. In other words, the facility itself is intrinsically unappealing.

But the causes of NIMBY go deeper. A fundamental cause of failure in the siting process is the absence of trust. Often, potential host communities do not trust the proponents of the siting process, nor do they have faith in the siting process itself. In many instances, they do not understand the need for the facility or its detailed engineering aspects. This is particularly true of waste incinerators, which are technically complex.

This distrust of authority is a phenomenon that pervades contemporary society. It is a natural reaction by individuals against big government and big business. It is an outwelling of feeling—a rebellion by the small, powerless individual against the faceless, powerful government or large corporation. After all, if these powers can't control the deficit or crime, why should they be trusted with this undertaking? There is often a feeling that the facility is being rammed down the throat of the local community.

Another important factor in the NIMBY phenomenon is a deep-rooted psychological trait: fear of the unknown. We live in a complex technological world. There are more than 45,000 different chemicals in production in our modern industrial society, and this bewildering array is very threatening to most people. They do not understand PCBs, dioxins, DDT, vinyl chloride, and the hundreds of other chemicals that are sensationally reported in headlines and are about to wind up in their back yard. It is a natural human tendency to be frightened and cautious of things we do not understand. The simple solution is to reject the facility and its Pandora's box of toxic chemicals.

The media and some activist groups are well aware of this aspect of human nature and use sensationalism and scare tactics to play on this fear. They describe in vivid detail the negative impacts of scenarios that are highly implausible or could be mitigated.

There are other factors that work against the siting of new waste management facilities. As human beings we inherently have an inertia, a natural propensity to continue to live our lives according to the routines that we have established. Thus, we are naturally opposed to intrusions that will cause disruption in our lives, even if their negative effects can be mitigated or compensated.

Spurred by distrust, fear, sensationalism, and disinformation, conflict almost inevitably arises in a potential host community. Such conflict is often very divisive and bitter. Frequently, even community members who support the facility may choose to drop out of the process simply to avoid the unpleasantness.

In summary, there are many valid reasons for the power and success of the NIMBY syndrome. In particular, the unattractiveness of waste facilities, combined with an inherent distrust of the proponent and fear of the unknown, can be fanned into a very emotional and divisive atmosphere which ultimately leads to failure of the process. This concept is illustrated in Figure 12.1.

Before discussing how these difficult hurdles might be overcome to create a win-win situation, let us look at two case histories of site selection and the NIMBY syndrome.

Radioactive Waste: NIMBY on the Grandest Scale

There are many instances in which the NIMBY syndrome has brought projects to a halt, but none can match the epic proportions of the search for low-level radioactive waste (LLRW) disposal sites in the United States. This exercise has been dragging on for the past two decades. The economic cost alone has been staggering, with hundreds of millions of dollars squandered in a fruitless effort to find disposal sites. It has been a nationwide drama, with a cast of

12.1   Conceptual view of two different paths for a siting process.

players that includes state governors, federal legislators, and movie stars. It is a fascinating case history complete with conflict, melo-drama, and pathos.

The stakes are enormous. Most people are not aware of how widespread is use of radioactive materials in our society, and how much our modern lifestyles and health depend on nuclear materials and technology. Every hospital in the world relies on nuclear tech-niques; approximately 150 million nuclear diagnostic procedures are performed every year. Virtually everyone either has undergone a nuclear medical procedure or has a friend or relative who has. The use of the radioactive element cobalt-60 to treat cancer patients has added a total of approximately 13 million years of human lifespan.

Radioactive techniques are also widely used in research. For example, radioisotopes are used in genetic research and in the DNA tests that are becoming common in police investigations. Radio-active materials are also widely used in industry.

Radioactive Materials and the Economy

A survey conducted in 1994 (Management Information Services, 1994) esti-mated that the direct and indirect impacts of the use of radioactive materials in the United States in 1991 (excluding nuclear power generation) were as follows:

- $257 billion in total industry sales
- 3.7 million jobs
- $45 billion in federal, state, and local government tax revenues

---

Like other industries, the production and use of radioactive materials produces wastes that must be safely disposed of. With this objective, Congress enacted the Low-Level Radioactive Waste Policy

Act of 1980, which laid on states the responsibility for disposing of LLRW generated within their borders. At the time, there were three disposal facilities operating in the United States, at Beatty, Nevada; Barnwell, South Carolina; and Hanford, Washington. To avoid a profusion of disposal facilities, the states were encouraged to group into entities called "compacts" that would share one disposal facility. It was envisaged that the repositories would be relatively few in number and that they would be situated within a reasonable distance of the places where the wastes are generated. (The search for a disposal site for high-level radioactive waste from nuclear power plants is not described here.)

Owing to lack of progress, the Low-Level Radioactive Waste Policy Amendments Act was passed in 1985. It established a number of target dates as well as penalties to ensure compliance. The year 1993 was the critical milestone by which the disposal facilities were to be operational.

Since 1985, the LLRW siting process has turned into a political quagmire, with huge sums of money spent in endless debate. Headway has been made in only one area, the formation of the following compacts (the states that are to host the disposal facilities are italicized):

- Appalachian: Delaware, Maryland, *Pennsylvania*, West Virginia
- Central Midwest: *Illinois*, Kentucky
- Central Interstate: Arkansas, Kansas, Louisiana, *Nebraska*, Oklahoma
- Midwest: Indiana, Iowa, Minnesota, Missouri, *Ohio*, Wisconsin
- Northeast: *Connecticut*, *New Jersey*
- Northwest: Alaska, Hawaii, Idaho, Montana, Oregon, Utah, *Washington*, Wyoming
- Rocky Mountain: *Colorado*, Nevada, New Mexico
- Southeast: Alabama, Florida, Georgia, Mississippi, *North Carolina*, Tennessee, Virginia
- Southwest: Arizona, *California*, North Dakota, South Dakota
- Texas: Maine, *Texas*, Vermont
- Unaligned: *District of Columbia*, *Massachusetts*, *Michigan*, *New Hampshire*, *New York*, *Rhode Island*, *Puerto Rico*, *South Carolina*

Although most of the compacts are organized along geographic lines, as would be expected to minimize transportation, there are also some strange bedfellows (see map, Figure 12.2). After serious

12.2   Alignment of state compacts; name of host state is underlined.

attempts at going it alone failed, Vermont and Maine joined in a compact with Texas. This arrangement could result in waste shipments crossing at least five states in the voyage from the northeast to Texas.

Not a single new disposal facility has yet come into operation, although the 1993 target for having operating facilities has long passed (Department of Energy, 1996). Several states have selected a preferred site and are seeking a license. Only California has a licensed site, but the process is mired in the transfer of land from federal to state jurisdiction. Several states such as New York and Michigan, have for all practical purposes given up searching for a disposal site. It is clear that the 1980 and 1985 acts have been abject failures.

Of the three repositories that were operating before the 1985 act, two remain open: Barnwell in the East and Hanford in the West. The disposal facility in Beatty, Nevada, was permanently closed in 1992. The facility in Hanford is now open only to members of the Northwest Compact and the Rocky Mountain Compact.

Barnwell, South Carolina

The repository operating in Barnwell, South Carolina, has an interesting and acrimonious history. In 1992, the Southeast Compact, recognizing that no state or compact would meet the 1993 milestone, authorized their facility to stay open for out-of-compact states until mid-1994. However, these outside states would have to show progress toward developing their own facilities and pay a surcharge of $883 per cubic meter ($25 per cubic foot) in addition to the $5,756 per cubic meter ($163 per cubic foot) already being charged to states outside the Southeast Compact.

On June 30, 1994, the Barnwell facility closed to out-of-compact states. This left 31 states and approximately 3,000 hospitals, companies, and institutions without a place to dispose of their low-level radioactive waste.

With no sign of a new disposal facility anywhere in the nation, South Carolina passed legislation to reopen Barnwell on July 1, 1995, to waste generators in all states outside the Northwest and Rocky Mountain compacts (which have access to the Hanford site), for a period of 10 years. In frustration with the lack of progress in its own compact, however, the law banned North Carolina from access to Barnwell.

When approval for this action was sought from the Southeast Commission, it was narrowly defeated. At the time this book was written, South Carolina had withdrawn from the Southeast Compact and was not permitting North Carolina access to Barnwell.

NIMBY in California

The Southwest Compact, with California as host, was considered to have the best chance of meeting the 1993 target. Even before entering the compact, California enacted legislation authorizing the Department of Health Services to site and license a disposal facility. The department hired US Ecology, Inc., to find a site and design, construct, and operate a facility.

Three candidate sites were identified in 1987, and Ward Valley was selected as the preferred site in 1988. This site is located in the Mojave Desert in a federally owned area with a very deep water table, low annual rainfall, and surficial geology of thick clay layers. After extensive studies, a site license application was submitted in 1989. In spite of numerous lawsuits and fierce opposition, a license for the facility was issued in September 1993—the first license for a new low-level radioactive waste disposal facility anywhere in the United States in 23 years. But the site is still not operating today.

The latest, and certainly the most puzzling, obstruction has come from the federal government, which initiated the entire compact process. The Department of Interior has for years delayed the transfer to California of the lands on which the Ward Valley facility is to be constructed.

The National Academy of Sciences studied the matter and issued a report in May 1995 which provided a qualified clean bill of health to the project. Issues considered included the potential for hydrological connection between the site and the Colorado River and the loss of habitat for the desert tortoise, an endangered species. After their report was issued, the Interior secretary agreed to the land transfer, but with some strong conditions, including continued federal involvement. The conditions were unacceptable to California.

There are many actors on the California stage. Greenpeace set up an encampment on the site in late 1995, vowing not to depart until the project was abandoned. But politicians, sensing that reputations and careers can be made on this emotional issue, are most prolific in this drama. They include county representatives, state land commissioners, and state and federal congresspersons. Even President Clinton entered the fray, vowing to veto any bill that approved the land transfer to California.

There is also squabbling among the proponents of the facility. US Ecology sued the state to recover some of the huge cost overruns caused by the continuing delay. The state has also been sued by the National Association of Cancer Patients and the California Radioactive Materials Management Forum. In turn, the state is suing the federal government over its refusal to transfer the land.

It appears that the radioactive waste saga in California will continue for a long time.

The Impact of NIMBY

What are the impacts of these long-running delays? The first is that the cost of disposal has become very expensive. For example, waste generators from outside the Southeast Compact were paying over $10,593 per cubic meter ($300 per cubic foot) in 1995 for access to the Barnwell facility (the disposal charge varies depending on the concentration and type of radioactivity). This compares to the 1986 cost of about $1,766 per cubic meter ($50 per cubic foot) and $25 per cubic meter ($0.70 per cubic foot) in the 1950s (Newberry, 1993).

The High Cost of No Forward Progress

The endless delays in establishing new low-level radioactive waste disposal facilities have not been without cost. It is estimated that a total of approximately $512 million in public and private funds has been expended on disposal facility development efforts since the passage of the 1980 Policy Act (Nuclear Energy Institute, 1996). This is a staggering price, considering that there has been virtually no progress made.

The high cost of disposal has, however, had a positive influence on how wastes are managed. Every effort has been made to reduce the amount of waste generated and, where possible, to use treatment techniques such as incineration and supercompaction to reduce volume. This has resulted in a significant reduction in the amount of nuclear waste that must be disposed.

Many generators are turning to on-site storage. For example, in California approximately 800 temporary storage sites have been set up at nuclear power plants, hospitals, universities, and research institutes. One has to wonder whether these temporary sites provide the same degree of safety that would be afforded at Ward Valley.

The use of nuclear medical procedures in many states is being seriously threatened, with a direct and significant impact on human lives. The use of radioactive materials in laboratories is being curtailed, threatening the position of the United States as a leading R&D nation. Industries are being forced to use inferior and generally more expensive techniques in their monitoring and processing. Some areas are faced with the prospect of entire industries having to relocate because of lack of access to disposal facilities.

What Happens without Disposal Capacity?

After Barnwell closed its doors to out-of-compact states, a survey was conducted of 680 companies and institutions that were denied disposal access, to determine what impact the closure was having on their operations (Organizations United, 1996).

- One year after loss of disposal access, 15% of the firms had cut products or services, resulting in increased costs and lower quality of life. Two percent of those surveyed felt that these cuts would lead to loss of life.
- Five medical institutions were forced to refer patients to other facilities.
- Of those surveyed, 59% incurred higher operating costs and 17% reported a loss of revenue.

Summary

The siting process has led to bitter fighting at all levels. State has fought state in desperately trying to avoid becoming the designated disposal host. States have fought the federal government in trying to have the 1980 and 1985 Acts overturned. But the most bitter skirmishing has been at the local level in states that have tried to site a disposal facility. Opposition groups have tried and generally succeeded in blocking progress by using a variety of obstructionist tactics, ranging from court actions to concerns about endangered species to physical violence.

Modern judicial, regulatory, and political systems place enormous power and responsibility at the grass-roots level of society. Consequently, opposition groups have had considerable influence in directing the siting process and the development of screening criteria and guidelines. The siting process has become extremely complex, and environmental assessments have gone to levels of detail never witnessed in the past. Thus, at every turn, the process has become more difficult, more complex and, most of all, more expensive.

NIMBY's influence now extends far beyond what might be considered unattractive facilities, and it is becoming more and more difficult to site almost any major project, no matter how beneficial to society it may be. It is difficult to envisage how society can continue to evolve without the ability to provide infrastructure for a growing population.

This raises a burning question: Are there siting processes that overcome NIMBY while protecting the environment and the rights of both local groups and the larger population, allowing society to flourish in a sustainable manner?

A Rare Success Story: Swan Hills, Alberta

Scattered in a sea of siting defeats are occasional, lonely islands of success. The most noteworthy occurred at Swan Hills, Alberta, where a hazardous waste disposal and treatment facility was sited in 1984 and commenced operation in 1987 (Henry & Runnals, 1989; Rabe, 1994a). Situated about 200 kilometers northwest of Edmonton, this was the only hazardous waste facility to be sited in North America during the 1980s. (The Swan Hills facility is described in chapter 11.)

At the outset, Alberta seemed an unlikely candidate to break the iron grip of NIMBY. The province began the 1980s with a private-sector approach to siting a hazardous waste facility. Two sites were proposed but were quickly rejected by fierce local opposition. Recognizing that this approach would be unlikely to succeed, the provincial government placed a moratorium on siting and established the Hazardous Waste Management Committee.

The committee proposed a siting process that involved the three main parties in what was then an innovative approach. One key was an emphasis on voluntarism: only communities that offered to host the site would be considered. The private sector's role was to plan and operate the facility. The province was responsible for developing siting criteria and a public information and involvement program. Particular emphasis was placed on communication, starting at the outset of the process.

In the early stages of the site-selection process, Alberta officials hosted more than 120 informational meetings throughout the province. Those communities that expressed interest in participation were offered additional information, including a detailed analysis of their geographic area. Fifty-two jurisdictions requested these analyses. Siting criteria were developed and applied through constraint mapping. The process emphasized consultation with the public.

Subsequently, 14 communities expressed interest in continuing in the site selection. Of these, nine dropped out, either because of technical unsuitability or in response to strong public opposition. The remaining five communities held plebiscites in 1982, and all five received overwhelming approval for hosting the facility.

Swan Hills, with 79% of its voters in support, was selected as the host for the facility in 1984. Situated 210 kilometers northwest of Edmonton and with a population of 2,400, Swan Hills was attractive because of its proximity to Edmonton, its good transportation links, and its relative isolation, so that the support of nearby towns was not required.

Acceptance of the facility by the residents of Swan Hills was probably helped by the competitive atmosphere that developed among the communities vying for the complex. The community leaders from the town of Ryley, for example, were outspoken in registering their disappointment at not being selected.

The strong support of local political leaders was essential in building public trust and grass-roots support. They emphasized the positive economic development potential, the voluntary nature of the exercise, the need for such a facility, and the concern that wastes in the province, including their region, were currently being managed in an unsafe manner. Regular informational meetings were held, and the mayor and council were proactive in encouraging citizens to attend the meetings. In particular, the meetings served as a useful forum to consider and refute claims from national and international environmental groups that the facility would pose a serious environmental threat to the area.

Economically, the construction of a $34 to $38 million complex, with 55 new permanent jobs, was attractive. A comprehensive package of benefits to the community also formed a key part of the siting negotiations. A grant of approximately $128,000 was provided to cover expenses of town meetings, consultations with experts, hiring a consultant to review monitoring data, purchasing a van to provide transportation of townspeople to the site, assistance in developing a golf course, and planting of trees. An important part of the package was a special medical surveillance program for all facility employees.

Since the Swan Hills Treatment Centre came into existence, the town of Swan Hills has enjoyed a period of relative prosperity. The facility has helped to overcome a decline in the oil and gas industry by creating 86 new jobs and also by luring new industries that wished to be next to a comprehensive waste management facility. The town has seen major increases in housing starts, a $3.8 million upgrade to the water supply, the opening of a modern hospital, the construction of a major new office complex, and the beginning of a major industrial park. An unexpected economic development has been the large number of technical tourists whom the facility attracts.

The facility and its operation are described in chapter 11. Recently it has had some financial difficulties, owing primarily to lower waste volumes than projected. Nevertheless, as a case study in the siting of a controversial facility, it is a remarkable achievement from which much can be learned.

Building an Equitable Siting Process

What is an equitable siting process? It is a process whose outcome is a win-win situation for all the stakeholders. An equitable process will find a site or sites with technically suitable characteristics, at the same time safeguarding the interests of the host community. It will ensure that the health of local citizens is protected, that their concerns are addressed, and that they are compensated for the inconvenience of hosting the facility on behalf of others. Furthermore, the process should generate an atmosphere that promotes calm negotiation rather than emotional divisiveness. An equitable siting process breaks down the walls of distrust and replaces confrontation with discussion.

Considerable effort has been expended over the past decade in trying to define an equitable siting process, and the following elements have emerged.

The public must understand and accept the need for the proposed facility. The need for the facility should be explained as part of a sensible, long-term waste management strategy, rather than in isolation.

A siting process should allow meaningful involvement by the local community. In fact, a voluntary siting process, including shared decision-making, is now advocated as the best approach. This effectively turns away from the "decide-announce-defend-litigate" pattern of the past to a "consult-decide-announce-consult-improve" process. It is important that the government of authority—local, state, or federal—and the proponent define the process in advance and make a commitment to using it. The local communities should be involved from the outset.

Economic benefits should be given to the host community to compensate for the inconvenience and potential risk they are accepting. The form of compensation can vary widely; it may include cash, tipping fees, jobs, and property value guarantees.

Stringent safety standards must be implemented to ensure protection of human and environmental health. This would include ongoing monitoring of potentially affected individuals as well as the ecological system. It is particularly important that the community be involved in defining this program so residents are convinced that their safety is being adequately protected.

Financial support can be given to the community at an early stage to hire its own independent expert consultants to review the proponent's studies and permit applications. This would alleviate concerns about "hidden" facts.

Local citizens should be given operational involvement in the process, including meaningful decision-making responsibility. Possible areas of involvement include overseeing the monitoring programs, membership on the management team, and a citizens' oversight committee—possibly with the power to shut down the facility should an emergency situation arise. Such involvement, more than any other factor, would show that the process is open and fair. The host community and its citizens should be a resource and a partner rather than an adversary.

A process should be used that is open-ended rather than a short or one-shot exercise. Presumably, involved parties will treat each more civilly if they know that they will be interacting over an extended period. It is also essential that both parties expect to achieve optimal results through cooperation.

Fear of the Unknown

Fear of the unknown is a powerful human emotion. Suppose that a stranger asks you to step into a pitch-black room. Naturally, you will feel very reluctant to enter. Not knowing what is inside, your mind fills the void by vividly imagining snakes, an elevator shaft, or any number of other frightening possibilities awaiting you.

Your worries would quickly dissipate if the light were switched on and you could see that the room was harmless—say, your own bedroom. Providing information to a community is like switching on the light in a dark room.

The siting process will depend on the specific characteristics of the community, including its demography, history, geology, and natural setting. The details of the type and size of facility will have to be carefully factored into the process. In other words, the siting process needs to be flexible so it can be tailored to the specific needs, concerns, and character of the community that is involved.

The Fly in the Soup

In recent years there has been a marked shift from a "technological" approach to a "voluntary" siting approach which incorporates many of the guidelines described above. Nevertheless, siting processes continue to come off the rails. For example, a Minnesota initiative to site a hazardous waste facility was modeled after the Swan Hills approach. Despite this, it ground to a halt (Rabe, 1994b). In the search for low-level radioactive waste disposal sites (described earlier in this chapter), some states have tried the voluntary siting approach—including very substantial benefit packages—but with no success.

Although the siting process has evolved significantly, it still has some fundamental flaws. What are the reasons for the continuing failure of siting processes?

The siting process is very sensitive and vulnerable, and tiny embers can easily be fanned into infernos of emotion. The process most often comes unstuck as a result of the involvement of activist groups who have only one goal: to disrupt the process. These groups are generally small and often are from outside the area. They have no desire to cooperate or negotiate; they feel the end justifies the means. Their strategy is to create distrust, bad will, and fear; and, for the reasons discussed in this chapter, it is not difficult to achieve these goals. Once a tense and disagreeable environment has been created, the community will often seek peace and quiet by withdrawing from the process.

The final, and currently missing, component is a mechanism which keeps the atmosphere calm, thus allowing objective dialog to take place. It is only in such a calm setting that the desires of the community can be considered and informed decisions made.

In the past, siting failures were often due to the insensitivity of the proponent and ramrod tactics. The voluntary siting approach has gone a long way toward remedying this problem. No such guidelines have been developed to ensure fair play by other groups.

The natural susceptibility of humans, particularly in a group, to emotional rather than rational behavior also contributes to the downfall of siting processes. It is akin to a buffalo herd which has been frightened into a stampede, controlled totally by emotion. For example, it is common at information meetings that activist groups organize their members into one area where they boo and disrupt proponent speakers or any community citizens who may take a stance that does not agree with theirs. This is very intimidating; it fosters a mob togetherness and virtually ensures that citizens will side with their view. It can create a very unpleasant atmosphere, and the easy way out for the townspeople is to withdraw from the siting process.

The Final Step

The ethics of tactics that raise emotions must be questioned. The creation of a hostile, emotionally charged atmosphere infringes on the right of community members to obtain information and make their own decisions.

A factor that is seldom discussed or considered is that trust works in both directions. Modern siting processes devote considerable effort to ensuring that the proponent deals with the host community in a fair and equitable manner. The final step in building an equitable siting process is ensuring that the opponents also act in a fair and equitable fashion—in other words, that there is fairness and respect in both directions.

Situations should not be allowed to develop where small but well-organized and vocal opposition groups use distorted facts and polemic to stampede the potential host community into dropping out of the siting process. Likewise, the proponent has no right to decide that a facility is going to be situated in a certain community, and then take steps effectively to railroad the facility through by coercion and other strong-arm tactics. An equitable process is one in which *both* the proponent and the groups representing the host community are fair and open.

It is essential that issues are thoroughly debated and that freedom of speech is not impaired. Nonetheless, false statements, distortions, and sensationalism should not be tolerated. The community has a responsibility to set rules that provide for factual debate without limiting anyone's right to free expression. It is not easy to formulate methods by which this might be done.

In summary, for meaningful siting processes, governing norms need to be established that guide the conduct of involved parties and allow for an objective, factual debate in a non-emotional setting (Rabe, 1992). These governing norms should not curtail debate or the freedom of speech, but they should offer a mechanism for filtering out unsubstantiated facts, falsehoods, and distortions.

Needless to say, methods of enforcing the norms will be even more difficult to develop than the norms themselves. Some possible approaches to developing an equitable siting process include the following.

A panel of citizens could be established to review the proponents' proposals and studies. Such citizen advisory groups (CAGs) have been used frequently in the past, with mixed results. Some refinements are recommended here. First, the community should seek assurance that the committee would be composed of fair-minded, respected individuals. The panel members should be selected and would be expected to act much like judges in a judicial system. That is, they would be expected to listen with open minds to all arguments and opinions and to balance them objectively; they would be expected to act as individuals and not represent any specific groups. Just as judges are sworn into office, panel members might be asked to sign an affidavit or take an oath that they will perform their duties in a fair and unbiased manner and act in the best interests of the community.

Second, the CAG could act as a conduit for information to the community and to the press. For example, public information sessions could be organized under the auspices of the CAG, who would establish guidelines for the speakers. They could also review and approve speakers.

Any advocacy groups, either pro or con, should clearly identify who they represent and who their members are, and they should accept responsibility for the factual content of their statements. All groups involved in the issue should undertake to behave by the established governing norms.

The community could contact the local media and request open, fair media coverage. Whenever items showing potential bias or distortion are submitted to the media, they should be researched to verify their content, and the other side should be given the opportunity to provide a counter-statement. Sensationalism and emotionalism should be avoided. This proposal would not be a curtailment of the freedom of the media; rather it would be a request for them to act in a professional and responsible manner.

Permanent new governmental conflict-resolution institutions could be established which have the expertise to develop compromise and resolve conflict in controversial siting situations. These agencies could assist the local CAGs and could mediate if difficulties developed. Acting as overseers or referees, they would speak out if they felt that any parties were not adhering to the governing norms. Needless to say, such agencies would need to be totally independent and have credibility to perform their duties.

Many delays in siting result from lawsuits and legal challenges. The proposed governmental conflict-resolution institutions might also be granted broader legal rights to adjudicate in such matters, rather than having them go before the courts.

These proposals do not guarantee that facilities will be successfully sited; communities may still choose to reject them, and they should have every right to do so. However, the proposed siting process ensures that decisions will be based on healthy and open dialog. In our complex technological world, we need such a new direction.

Tips from the Trenches

From 1987 to 1995, the Canadian government developed and implemented a method to find a disposal site for a large volume of historic low-level radioactive waste which arose from the refining of uranium ore. Called the Cooperative Siting Process, it is the most innovative and ambitious siting program undertaken in Canada. Principles and guidelines were established for the process at the outset to ensure that potential host municipalities would participate on a voluntary and partnership basis. The process was designed to be open, participatory, and democratic. Extensive consultations were conducted starting with invitations to all 850 Ontario municipalities, followed by detailed discussions with 26 municipalities that decided to become involved. Although most of the initial volunteer sites opted out over the ensuing eight years, eventually one community volunteered a site. Unfortunately, largely because of costs, the government has to date not proceeded with developing the facility. The process was often acrimonious and had many hitches along the way. Here are some of the practical pointers that emerged (Lafferty, 1996).

Community Liaison Groups were established at each potential host municipality. Difficulties were frequently encountered with these groups because they were seen as infringing on the role of the elected councils. It is felt that such groups would be more productive if they were appointed by and reporting to the elected representatives as, say, an advisory committee. An orderly mechanism for regularly replacing representatives would provide continuity as well as renewed energy when the process runs over several years.

The process was vulnerable to media coverage of the negative, the alarming, and the eccentric, but seldom the factual. Perceptions of the public and politicians were often formed by these media reports. It is important to devise methods to obtain unbiased media coverage.

Open houses are an effective means of communication. They should be designed so that they provide information, are honest, lack hard-sell pressure, and avoid presenting a platform for lobbyists.

The perceptions of nontechnical people about the safety of the proposed facility were not swayed by costly technical assessments. Far more effective are simple analogies and demonstrations that explain in a hands-on manner. For example, visits to disposal facilities that had been operating safely for years did more to instill confidence in the merits of the proposed facility than did any technical study. Concrete examples, rather than abstract studies, are the key to changing perceptions.

Discussion Topics and Assignments

1. To which compact does your state belong for developing a low-level radioactive waste disposal facility? Contact the lead organization and find out what progress has been made. What are the main difficulties they have encountered, and how do they propose to overcome these?

2. Have you observed any examples of the NIMBY syndrome in your community? What were the main issues? Do you feel that a fair outcome resulted?

3. If you were chairing an open-house meeting for siting a waste disposal facility and one opposition group became noisy, rude, and disruptive, how would you handle the situation?

13

A NEW APPROACH

We need waste disposal methods that allow the human race to live on this planet in harmony with nature, preserving our resources and habitat and leaving a legacy for our children and grandchildren that does not deprive them of opportunities. These changes will not come easily; they will require resolve and foresight.

Starting from Basics

Just as a mathematician develops the proof to a mathematical theorem, we must start from a basic axiom, and step by step, following a logical progression, we must build a practical framework for waste management. We started this task in chapter 2, where we derived three general principles from the axiom of sustainable development. Can we apply these general principles to develop practical guidelines—first, to overcome the shortcomings of existing landfills, and second, to find other, innovative disposal methods that will conform with sustainable development? Let us look at each of the three principles in turn.

The Health and Environment Principle

Human health and the environment must be protected, both now and in the future. This principle is fundamental and places important constraints on the siting and design of disposal facilities, and also on the form of the waste. In particular, the final four words, "and in the future," are very important.

245

This principle can be satisfied in two ways: by reducing the toxicity of the wastes so they pose minimal risk, or by containing wastes so that they cannot escape and cause harm. In some cases, the latter method includes controlled leakage at a rate that the environment can assimilate without long-term degradation.

The Future Generations Principle

Wastes must be managed so that no burden is placed on future generations and they are not deprived of the opportunities we have had. In other words, our grandchildren should not have to spend their valuable resources to solve our waste problems, nor should they be denied resources because our generation has depleted them. Neither should their health and environment be placed at risk because of our actions.

The main impacts of landfills on future generations are (1) the requirement to provide ongoing guardianship and maintenance; (2) the loss of valuable land; and (3) impairment of groundwater, surface water, and the atmosphere. There are essentially only two ways the first and third impact can be prevented. First, the volume and toxicity of the wastes can be reduced so that they pose no threat. Alternatively, the wastes can be placed into a containment that is isolated from degradational forces so that it does not require maintenance, and that is sufficiently secure that wastes will not escape in future. The second impact is discussed under the next principle.

Conservation Principle

Nonrenewable resources must be conserved. Although this principle seems simple, its application leads to a number of important guidelines. Three resources are affected by waste disposal:

- Land used in developing landfills
- Groundwater contaminated by landfill leachate
- Resources bound up in the waste itself

Let us review the shortcomings of landfills:

- Many older landfills, generally constructed without liners in areas selected for convenience, are leaking leachate into groundwater systems.
- Newer landfills which incorporate modern engineering barriers will leak eventually, probably within 50 to 100 years of closure.

- Landfills emit gases that can cause both local and global environmental problems.
- Landfills contain valuable resources such as paper, metals, and plastics that should be recovered for recycling and reuse, as well as nonrecoverable materials that can be incinerated to generate energy; thus, landfills are a resource.
- Many landfills are not collecting landfill gases—a waste of a valuable energy resource.
- Landfills, particularly modern ones, protrude well above the surrounding landscape and are exposed to erosion; thus, they will require monitoring and ongoing maintenance for centuries to come.
- Landfills occupy large tracts of land that could be put to better use by society.

In summary, landfills generally provide only short-term security. When the long term is considered, they violate all three basic principles necessary to achieve sustainable development.

Guidelines

This book argues that we should decrease our dependence on near-surface landfills to the maximum degree possible. Although this is not feasible in the short term, we should set the goal of eliminating near-surface landfills altogether over the long term. But how do we do this in a manner that is consistent with the three guiding principles? Guidelines can be derived from sustainable development and the three basic principles; this concept is illustrated in Figure 13.1.

Recycling

A direct consequence of the conservation principle, and underpinning both of the other main principles, is the vigorous application of the "three Rs" to ensure that the amount of wastes that ultimately need to be disposed is minimized. In other words, recycling will not only greatly reduce our need for landfills; it will also reduce the demand for energy, reduce pollution of land, water, and air, and conserve valuable raw materials.

The importance of reducing waste quantities cannot be stressed sufficiently. It is the key that will unlock the door to the goal of sustainable development. The three Rs must form a fundamental pillar of any integrated waste management program.

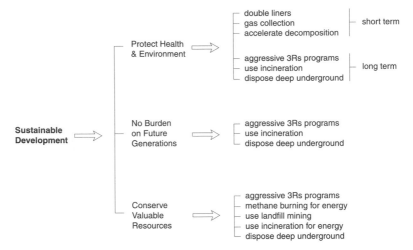

13.1   Development of waste management guidelines based on sustainable development.

Incineration

Even with vigorous recycling—as the Japanese example shows—a very large amount of waste will still require disposal. The only other technology available today to reduce it is incineration. It destroys organic compounds, reduces waste volumes, and renders the waste into an ash that is amenable for treatments such as solidification. These results support the health and environment principle. Electricity or district heating can also be generated from waste, preserving valuable resources such as oil, coal, and natural gas for future generations. Incineration has the potential to become an important and necessary complement to recycling.

Reduce Toxicity

Finally, even after recycling and incineration, some wastes will remain. All three principles would be satisfied if these could be rendered harmless prior to disposal. This is difficult to achieve completely, however, because it requires that the waste undergo chemical transformation into nontoxic elements. This is achievable for organic compounds, but not for inorganics. Inorganics should be put into an inert, leach-resistant form—for example, encapsulated into cement as is done at the Swedish nuclear repository discussed in chapter 11. Additional barriers could also be incorporated, if deemed necessary.

Breaking Dependence on Near-Surface Landfills

As noted earlier, all three sustainable development principles suggest that we decrease our reliance on near-surface landfills. The
following discussion identifies both short-term and long-term strategies to do this.

Short Term

In the short term, landfills will continue to be a necessary part of
waste management systems. During this period, they should be
designed and operated in a manner that satisfies the three principles
to the extent possible. This implies several requirements.

First, landfills should incorporate double liners and leak detection systems; contingency plans should be developed to deal with
leaks. Second, landfills should be sited carefully in areas that provide natural protection and assimilation of contaminants, should
leakage occur. Third, comprehensive gas-collection systems should
be used to ensure that toxic organic gases are not released to the
atmosphere. Energy should be produced from the methane.

Finally, short-term decomposition of waste should be accelerated
by using the bioreactor concept while liners and barriers are intact.
The anaerobic conditions that favor decomposition can be obtained
by using covers that permit infiltration, by adding sludge and water,
and by recirculating leachate. In other words, immediately upon
closure a landfill should be encouraged to "cook" its contents vigorously for a number of years. Once the greater part of decomposition has occurred, an impermeable cover can be placed over the
landfill, the leachate pumped out, and the landfill left in a dry entombed state.

Long Term

The short-term agenda consists of steps that are relatively common-
place; many of them are already being implemented by progressive
waste managers. In the long term, more radical measures are necessary. Some potential approaches are discussed here.

Landfills could be mined to recover recyclable materials and
energy, as described in chapter 8. This method would greatly extend the lifetime of landfills thus reducing the need for new ones.
Furthermore, older, closed landfills, particularly those that pose an

environmental hazard, could either be reclaimed for other beneficial uses or upgraded in terms of environmental protection.

Because inorganic compounds cannot be destroyed, secure long-
term containment is necessary. In simple terms, either disposal facilities should not leak at all, or any future leakage should be at such
low rates that it does not harm the biosphere. This containment
should stay in place over the period that the wastes remain hazardous. For inorganic wastes, this is essentially in perpetuity.

Once institutional controls have ceased, inadvertent intrusion
becomes the primary risk for surface facilities (Eedy and Hart, 1988).
To prevent people from unintentionally building on or digging into
a closed disposal facility, it should be isolated—removed from areas
that humans typically inhabit.

Chapter 8 explored a broad spectrum of waste disposal methods.
On closer scrutiny, most of these approaches have some fatal weaknesses. However, one alternative—deep underground disposal—
appears to have potential. In this method, wastes are placed deep
underground where sturdy rock formations that have been undisturbed for millions of years will protect the wastes for similar periods. This appears to be the only method that can economically provide both long-term containment and isolation.

The selection of a proper geologic host site should be done carefully to ensure that groundwater is protected. For example, silicate
rocks such as thick clays or granitic formations have very low solu-
bilities, and when few fractures are present they can be highly impervious to water passage. Some formations, such as sands and
porous limestones, are not suitable and should be avoided.

Waste should be emplaced as deep below the surface as practically possible. Depth not only provides isolation from the elements
and protection from future human intrusion, but it also places the
waste into zones that are largely impermeable and that contain very
small amounts of slowly moving groundwater. Groundwaters at
depth are often saline owing to the long time they have spent in the
subsurface.

Underground space is a vast domain which humans have explored and exploited only minimally, and thus it is suitable for
waste disposal. This is not a new concept: the nuclear industry has
been exploring deep disposal of its wastes for many years. An impediment to extending this method to municipal solid waste is cost;
it would be economically feasible only if waste volumes were greatly
reduced. Incinerator ash (particularly fly ash) and hazardous wastes
would be prime candidates for underground disposal.

Change in Emphasis—A New Waste Management Hierarchy

As described in chapter 4, most jurisdictions in North America have developed an integrated waste management policy with the following hierarchy of methods, from most important to least important:

- Reduction at source
- Recycling
- Incineration
- Landfill disposal

In practice, though, this hierarchy is often inverted. Although reduction is at the top of the list, little is actually being done, with most efforts focused instead on recycling. For example, very few people use their own reusable bags for shopping. There has also been a trend away from using refillable beverage containers to once-through plastic or aluminum containers. Incineration is supposed to take preference over landfilling, but in some jurisdictions—such as California and Canada—it is not being considered because of public opposition. Landfilling forms the main part of the overall system (see Fig. 4.2). In practice, the current waste management hierarchy is:

- Landfill disposal
- Recycling
- Reduction at source
- Incineration

We propose a revised waste management hierarchy that meets the three principles supporting sustainable development:

- Reduction at source
- Recycling and incineration
- Phase out near-surface landfills

Currently the United States and Canada produce more garbage per capita than any other countries. Action is required to reduce significantly the amount of waste produced at source. This is not an easy task, because it requires a fundamental change in attitude. We must change our lifestyle and place less emphasis on consumerism. Source reduction is to waste management what preventive medicine is to health.

Human beings have risen to challenges before: we have placed humans on the moon, climbed Mount Everest, found cures for diseases, even transplanted hearts. The lifestyle change associated with implementing a comprehensive three-Rs ethic is a similar challenge.

Recycling is a vital component of solid waste management. It is essential that society place much greater importance on recycling and develop the ethic and infrastructure to support it. This has been recognized, and there has been a rapid increase in recycling and composting programs since the mid-1980s. The total amount of waste going to landfill, however, has decreased only slightly (see Fig. 4.2).

Significantly more diversion of waste from landfills must be achieved. Of all the technologies that are currently available, only incineration is capable of fulfilling this objective.

In the proposed hierarchy, incineration would function together with and complementary to recycling. It would not compete with recycling but instead would deal only with those wastes that cannot be economically or practically recycled. Incineration would recover the energy content of wastes that cannot be recycled or composted. Incinerators, with energy production, would be complements to recycling, not competitors. Both are needed, and both are essential.

The proposed waste management hierarchy places near-surface landfills at the bottom of the list; furthermore, it recommends that proactive steps be taken to decrease landfill use in the near term, and that they be phased out completely in the long term. Closed landfills should be rehabilitated. New, radically different approaches—such as landfill mining, underground disposal, and beneficial uses for incinerator ash—are necessary to achieve this objective.

Non-Technical Issues

There are many hurdles to overcome in transforming our society's philosophy from convenience and unbridled consumerism to sustainable development. Although the solutions to the waste crisis are essentially technical, they cannot be implemented without major efforts in the social and regulatory arenas. There is a major non-technical component in solving the waste crisis.

First and foremost, attitudes must change. A united front is necessary to reduce the amount of waste we create. To accomplish such a transformation, everyone must get involved—not just waste management experts. There is a need for education, fiscal policies and taxes to encourage waste reduction, and penalties for using land for disposal. Everyone needs to be a player. We must recognize that poor waste management practices are just as devastating to our future as clear-cutting old-growth forests, or depleting fish stocks, or using up oil resources.

There is also a need for change in waste management technologies. The current situation is reminiscent of the time when the automobile began to replace the horse and buggy. People questioned the need for endless kilometers of paved highways and wondered whether it would be possible to build gas stations at thousands of convenient locations. Objections were raised about the unknown health effects of traveling at speeds as high as 30 kilometers per hour.

Waste management faces a similar challenge in introducing new technologies such as incinerators and underground disposal. Communities need to be persuaded to accept these facilities. Environmental assessments and public hearings will need to be held, and there will be the inevitable NIMBY arguments.

It will take some time to make this transition. What can be done to resolve the non-technical issues?

A major impediment in solving the waste crisis is the lack of general knowledge about wastes and waste issues. For example, few people understand the health effects of toxic compounds such as PCBs and pesticides, or realize that air emissions from landfills are as bad as or worse than those from incinerators.

We live in a complex world which does not readily lend itself to analysis in catchy one-liners. If one thread is disturbed, repercussions will be felt somewhere else in the vast, interconnected ecosystem web. To make sensible judgments, it is necessary to understand complex waste issues. This can come only through education and experience. Thus, it is important that information programs explaining the waste crisis, the hazards associated with wastes, and alternative technologies be initiated through school systems and the media.

Knowledgeable, well-informed politicians are essential in bringing together stakeholders and reaching effective, rational decisions. Thus, efforts must be made to inform and involve politicians in waste management issues. Since financial issues are always a major concern, politicians should be presented with detailed information on costs of relevant alternatives.

Different waste types, ranging from municipal to industrial to radioactive wastes, should be dealt with in a logical and consistent manner. As discussed in chapter 6, the hazards posed by these wastes are similar, varying only in degree. Yet these waste types are generally regulated by different agencies which often set widely different standards.

Regulations should also be made uniform across different juris-
dictional boundaries—municipal, state, federal, or international.
Much must be done to make laws and regulations and guidelines
consistent with the principles espoused above; they need a rational basis. Currently, too many regulations have been developed that
address a specific waste in isolation from other waste types and
other hazards that society faces. They are often based on emotional
arguments rather than on a solid foundation of science. Regulations have, on more than a few occasions, been established to serve
political purposes rather than providing real protection of the
environment.

There is also considerable fragmentation of waste management
responsibilities among different levels of government, often with a
lack of coordination or sharing of common goals. For example,
wastes are often shipped from one jurisdiction to another because
of cheaper disposal costs or less rigid regulatory requirements.

The current difficulties in siting waste management facilities (the
NIMBY syndrome) need to be resolved in fair and equitable fashion. On the one hand, we must be able to protect ourselves against
environmental threats. On the other hand, we need to be able to site
and build disposal facilities. By not taking action, we are placing
the environment at greater risk.

We have presented fundamental principles to guide the selection
and design of waste management facilities. Based on sustainable
development rather than convenience, these principles should help
in developing directions for resolving the waste crisis. In addition,
we presented specific suggestions for overcoming the emotion,
antagonism, and distrust that invariably arise at siting processes,
and replacing them with a calmer, more thoughtful negotiation.

In summary, a concerted effort is needed by waste engineers,
politicians, lawmakers, teachers, mediators, and—most important—
all the rest of us. Through informed discussion, attitudes toward
waste generation will change over time, and we will be able to
implement solutions to the crisis.

Discussion Topics and Assignments

1. This chapter argues that landfills are not good for the environment and should be phased out. Do you agree or disagree? Why?

2. How do you think that landfills could be phased out, and what period of time would be required?
3. Many people are opposed to incinerators. If they are not to be used, what are the alternatives?
4 Develop several scenarios showing how intentional and unintentional intrusion could compromise the integrity of closed landfills. For each scenario, describe how human health and the environment would be harmed.
5. Do you agree with the proposed waste management hierarchy? Why? How would you change it?

14

FUTURISTIC GARBOLOGY
A Vision

We gaze with some apprehension at a sleek, shiny machine that looks like a cross between a sports car and a small spaceship. Illuminated in a vertical cylinder of light, the time capsule silently awaits our entry; it gives no clue to what we can expect at our destination, the year 2032.

With racing hearts, our small group of garbologists enters; we strap ourselves in. Soon the countdown begins. There is only the briefest feeling of levitation, a slight rising sensation in the pit of the stomach as though descending in a fast elevator, and then we are there.

As we exit, our curiosity is at fever pitch. What has happened to the Earth in the three decades we miraculously skipped over? Before we departed, the world's population was rapidly approaching 6 billion, with many signs that the environment was finally wilting under this onslaught. We could only guess at what three more decades of continued environmental degradation might yield. Would we find air that was breathable, only traces of an ozone layer, any remnants of tropical or old-growth forests, any parks or green spaces in cities? Would North America be one giant parking lot?

With these questions buzzing through our brains, we begin our futuristic exploration, like archeologists working in reverse. We move invisibly through this new time domain, knowing that we can only observe and not affect anything we see.

It is clear that we are still in New York City, but what a change! The streets are no longer choked with car traffic, although small motorized bicycles, some built for two or three people, are darting everywhere. There is no haze in the sky, the air feels clean and brisk, and the streets are completely free of litter. Recycle containers are ubiquitous; they are green and divided into six compartments. As we soon discover, every house, apartment building, streetcorner, park, airport, shopping mall, and baseball diamond has recycling containers; there are no waste bins. People treat garbage as a resource rather than as something undesirable, and they spend considerable effort in separating the various recyclable components, whether they are at home, at work, in a cafeteria, or at play.

As we look around and see the future, we are pleased. There has been a revolution in the attitude of the public, and an ethic of conservation and the three Rs now pervades all levels of society.

This major change in mentality did not come easily. Plugged into the virtual-reality computers in the public library, we witness the environmental crisis of the previous decade. Growing populations, an avaricious consumer society, and the expanding economies of Third World countries led to enormous stress on natural resources as well as on the environment. For decades, even centuries, doomsayers had warned that the Earth was finite, that it could not support unconstrained population growth, and that disaster was imminent. For years, these Malthusian pessimists were derided and ignored.

But finally, in the 2020s, their predictions started to come true. Power blackouts began to hit with frightening regularity. The price of consumer goods skyrocketed owing to shortages of basic materials. There were riots at stores and malls. Curfews were imposed in large cities to limit exposure to the dank, health-sapping smog. Thanks to the library's virtual-reality computer systems, we are not just watching the news; we feel like we are living in the middle of the nightmare. But although it was a black period in history, it did the trick. People finally realized that action, rather than words, was needed.

The renaissance was like a wildflower blossoming among the cinders of a burned-out forest. Conservation fever suddenly gripped the world, the like of which had never been experienced before, except perhaps during the great wars of the twentieth century. In a single decade, society completely overturned the consumerist lifestyle that had been practiced for so long—and probably saved the Earth by doing so.

Emerging from the library, we observe that conservation and waste reduction touch the life of every individual. For example, stores do not provide shoppers with "free" bags. It is an automatic reflex to bring reusable bags when shopping, and the bags are made of remarkable new ultrathin materials that fold into the size of a matchbox so they can be conveniently carried in a pocket or purse. Consumer pressure has caused packaging of goods to become more restrained. Shrink-wrap and cardboard packaging are only infrequently seen, when there are no alternatives. Bulk goods are prevalent in everything from food to nuts and bolts at the hardware store. Individual packaging of small items is not quite extinct, but it is fading.

We observe a dramatic shift away from single-use, disposable products to more durable multi-use products. An important regulation was instituted in 2029 requiring that all new durable goods carry a sizable refundable deposit to ensure that the item is returned to the manufacturer for recycling. This includes things such as refillable drink bottles, whence the idea came, to computers, tires, lightbulbs, and photocopiers. The automobile industry perhaps best symbolizes the change of mindset. No longer do they change the design of their cars each year. Instead, like the Volkswagen Beetle of our era, the models retain the same form for many years, with the emphasis on durability and practical function rather than on glitzy style.

New companies have been created that collect broken and discarded household appliances, including electronic goods such as televisions, power tools, and furniture, for refurbishing and reselling. These firms have flourished, some with franchises across the country. Shares in these companies have been the darlings of the stock market for the past few years.

Although virtual-reality computers dominate all facets of life, especially education and entertainment, society still produces real, not virtual, garbage. It is recognized that placing this refuse into landfills poses an environmental risk. Government programs have been established to exhume old landfills systematically and reclaim them, eliminating the risk of groundwater pollution and allowing the land to be put to other uses. These projects extract useful items such as metals and glass, with the rest generally being incinerated. A number of landfills have begun recovering paper for pulp and paper plants, using a new technology that separates paper and cellulose products from mixed waste.

The public is engrossed by the archeological aspects of these landfill mining programs and is regularly entertained by fascinating insights into the lives of their ancestors. People seem particularly amused by the things that were discarded in the past—items which for years now have been illegal to place into landfills. We are embarrassed, recognizing that it is our refuse that is the subject of their amusement.

Collection of landfill gas has become mandatory, and all remaining landfills, mostly older ones, have comprehensive gas collection systems. Although the prime purpose is to prevent gases from polluting the atmosphere, the gas also provides valuable energy. Most of the sites generate electricity, but a number sell directly to nearby industrial or commercial users who fire their boilers with the gas. Some landfills upgrade the gas to pipeline quality and sell it to natural gas utilities which add it to their pipelines for distribution. It is estimated that landfills in North America contribute over 5,000 megawatts of energy, saving a considerable amount of fossil fuel and the attendant emissions to atmosphere.

With the smaller amounts of garbage, it is now practical to transport waste over long distances. As a consequence, landfill sites are regional in scope, serving many cities rather than individual municipalities as in the past. There are fewer landfills than at any point in the past century. Rail delivery of waste has become routine, and the landfills have specialized equipment to unload the railcars and deliver the waste to the operating cells.

Many regions are turning to deep underground burial of their wastes. This has been a boon for areas, such as the northeastern United States, where land space is at a premium. Waste is received at surface transfer stations and carried by conveyors to waste shafts where it is dumped into a series of large caverns about 150 meters below ground surface. Mining machinery moves, places, and compacts the garbage. Once the caverns are filled, the shafts and other underground openings are sealed with concrete.

Rock aggregate produced by the mining of the underground caverns is used in road-building and other civil engineering projects. This in turn has added cement-making, aggregate, and other spinoff businesses to the already burgeoning industrial parks situated over the top of these landfills. The cement and aggregate are also used to encapsulate some of the wastes, particularly incinerator ash, into a durable, rocklike form prior to placement underground. In addition, research has resulted in new applications for incinerator ash, including road construction, concrete piers, and other marine structures.

Although much less waste is going to landfill, energy generation from methane has not declined appreciably, owing to the efficiency with which the gas can be collected from subsurface facilities. The major benefit of going into inner space, however, has been safety.

No longer are there complaints of groundwater pollution or bad odors associated with landfills.

The few remaining aboveground landfills, generally the older ones, operate as bioreactors in a fast-cooker mode. Every effort is made to promote rapid anaerobic decomposition of the waste. Considerable research on the biochemistry of anaerobic digestion has been done in the past decade, and this has led to major advances in the science and engineering of landfills. It is now possible largely to eliminate organics in less than 20 years, enhancing the production of methane during this period. Leachate is recirculated; none is sent to sewers. Double and even triple bottom liners are standard to prevent any leakage, and permeable or semipermeable covers are used to allow the controlled entry of water into the landfill during this period. Once the bioreactor has effectively cooked the organic contents, the permeable covers are replaced by watertight caps, the leachate is pumped out, and the landfill is left in a dry entombed state, awaiting its turn to be mined.

With landfills being phased out, the material recycle facility has become the mainstay of the waste management system. All waste goes initially to the material recycle facility, where recyclables are separated and packaged for sale. We are fascinated by the technologies that have been developed to separate different recyclable materials. One machine separates the different plastics by using barcode scanner technology to read the plastic symbols. Another machine separates glass by color as well as by size of fragments. Everything has become highly mechanized, and very little scrap is left over. The final products are baled and packaged as they come off the sorting lines to the exact size and specification required by customers.

Processing recyclable materials has become a major growth industry. The brightest, most creative young men and women choose careers in developing new recycling technologies and designing new products from recycled materials.

An interesting development is that industrial parks have sprouted up around the major waste management facilities. Dozens of small industries have attached themselves to recycle centres and landfill mining projects in a symbiotic relationship that furthers the conservation ethic. These companies are very innovative and entrepreneurial in recovering, refurbishing, and reselling every conceivable type of material, including car parts, white goods such as refrigerators and stoves, scrap metal, small appliances, and electronics. Small manufacturing industries produce a large variety of products using recycled materials to decrease their costs and give them a competitive edge.

These "scavenger" industries, in concert with the very thorough recycling programs and incineration, ensure that only a very minimal amount of waste actually goes to disposal. Energy production, either from burning methane or from waste incineration, has also been a contributing factor in the growth of these industrial parks. There are jobs in working at the power plants, and the availability of cheap power has attracted industries that can use landfill gas either in their processes—for example, to manufacture carbon dioxide—or to fire their boilers, or to supplement their own natural-gas-fired cogeneration systems.

We are amazed at how the garbage business has escaped from yesteryear's stigma of being dirty, smelly, and unwanted, to an association with industry, jobs, and opportunity. The industrial-park aspect of recycle centers and landfill mining projects has been a significant contributor to overcoming the NIMBY syndrome, and now communities compete to attract these industrial centers.

Composting facilities are integral parts of these centers, using primarily in-vessel and enclosed systems. By law, every home must have a composter. The ample supply of high-quality compost has led to an increase in vegetable and flower gardens, which has had a beneficial effect on health through improved diet. In addition, public parks are flourishing, and every city we visit is attractively decorated with flowers and plants. There is an underlying sense of civic pride that is sadly missing from our own era.

Reduction in waste volume is considered an integral part of properly managing waste. For this reason, incineration is an important component of both the landfill mining programs and recycle centers. Instead of being ugly objects of scorn, as in the past, many incinerators now are the heart of community centers, surrounded by swimming pools and greenhouses that are heated by the energy they produce. These centers are emblazoned with colorful gardens nurtured by compost from adjoining composting facilities. Because of ongoing concern for human health and the atmosphere, pollution control technology has evolved considerably and air emissions from incinerators have decreased substantially. Nevertheless, in response to the public's seemingly unlimited appetite for environmental information, electronic billboards are installed near incinerators and display instantaneous readings of stack emissions for all to see.

By law, all incinerators must generate useful energy, either as district heat, process heat, or electricity. It is estimated that more than 15,000 megawatts of electricity are generated annually by waste incineration in North America. Combined with energy from landfill methane, this not only saves renewable resources such as oil but also significantly reduces emissions of greenhouse and acid gases into the atmosphere.

A major breakthrough has been the development of beneficial uses for incinerator ash. New epoxy technologies combined with engineering developments in concrete have led to use of ash from garbage incinerators and also from coal-fired power stations and other ash-generating industries, as a construction material. This has not only greatly reduced the amount of material going to landfill but has also decreased the demand for aggregate, which was becoming very expensive owing to dwindling supplies, and thus has helped preserve natural areas from quarrying.

We learn that the three Rs ethic affects not just garbage, but many other areas as well. For example, the consumption of all natural resources—water, electricity, natural gas, petroleum products, and more—has decreased noticeably. The dire predictions of energy shortages have been offset in large part by the "garbage electricity" from methane and incineration, in conjunction with this conservation ethic. The per capita demand for electricity and other energy sources has declined to approximately 55% of what it was in our era. Many small non-utility generating stations, based mostly on hydro, solar, and wind, have sprung up and make a substantial contribution to the national electricity grid.

Just before we return to our own time, we catch a late-breaking news story announcing that in the past year, not only did the amount of refuse generated decrease to 0.7 kg per person (about 35% of what was being produced in 1998), but 93% of this was diverted from landfills.

A sense of melancholy settles over us as the transporter brings us back through the decades to the present. Although the future we have seen is a good one, we wonder whether it was only a dream, a small piece of virtual-reality fiction. We also muse on whether it would be possible to instill the futuristic conservation ethic in today's society, before the globe is thrust into a dark abyss of crisis. Certainly, almost all the waste management technologies we observed in the future are within our reach today. But, sadly, one essential element is missing: our own resolve and determination.

Discussion Topics and Assignments

1. Do you think that the world will be plunged into an environmental and resource crisis in the future? Describe what you think will happen, assuming a worst-case scenario.
2. What can be done to instil a "futuristic conservation ethic" in today's society?
3. Describe how you think a more environmentally conscientious future society would manage its waste. What parts of the scenario presented in this chapter do you agree or disagree with?

GLOSSARY

Acid gases: primarily sulfur dioxide, hydrogen chloride, and nitrogen oxides emitted by waste incinerators and fossil-fuel power plants.

Aerobic: requiring oxygen (in contrast to anaerobic).

Alpha particle: a positively charged particle consisting of two neutrons and two protons emitted from the nuclei of radioactive elements.

Anaerobic: not requiring oxygen (in contrast to aerobic).

Anesthetic/narcotic: consciousness-attacking.

Bacteria: single-celled microorganisms lacking chlorophyll, some of which perform useful functions in the human body. Diseases such as diphtheria, tetanus, and botulism are caused by toxin-producing bacteria.

Baghouse: a pollution control device used with incinerators, in which particulate matter in exhaust gases is captured by a series of fabric bags.

Bentonite: a type of clay mineral that expands when moist.

Beta particle: electron emitted from the nuclei of some radioactive elements.

Biodegradable: material that can be broken down or decomposed into simpler substances by bacteria. Paper and most organic wastes are biodegradable.

Biomedical waste: waste containing toxic compounds associated with living cells.

BOD: Biological Oxygen Demand, the amount of dissolved oxygen needed by aerobic decomposers to break down the organic materials in a given volume of water over a specified period of time. It is a measure of the amount of biodegradable organic material in the fluid being measured.

Carcinogenic: cancer-causing.

Clay: extremely fine-grained sediment consisting of hydrous silicate minerals.

COD: Chemical Oxygen Demand, the amount of oxygen consumed (in parts per million) in the oxidation of organic and oxidizable inorganic matter in industrial waste water.

Composting: a method of solid waste management in which the organic component is biologically decomposed under controlled conditions to a state where it can be used as a soil amendment without adverse environmental impact.

DNA: deoxyribonucleic acid, the long-chained double-helix molecules in the nuclei of living cells that contain genetic information.

EPA: U.S. Environmental Protection Agency, responsible for managing federal efforts to control air and water pollution, radiation and pesticide hazards, ecological research, and solid waste disposal.

Epidemiology: the study of the causes of diseases among a population.

Fission products: isotopes created when a larger element, usually uranium-235, is split apart. They generally have relatively short decay times of seconds to decades.

Flocculation: process whereby suspended colloidal particles combine to form clumps, usually implemented so that they can be filtered or settled out of water.

Gamma ray: electromagnetic radiation emitted by the nuclei of some radioactive elements.

Geogrids: plastics with a very open, strong, gridlike structure (i.e., large aperture), used to provide reinforcement.

Geomembranes: thin sheets of impervious rubber or plastic used as liquid or vapor barriers.

Geonets: parallel sets of polymeric ribs at acute angles to each other, used to provide drainage for fluids.

Geotextiles: similar to regular textiles but made of synthetic fibres rather than natural ones. Flexible and porous, they are generally used to separate materials of different particle size, for example, to prevent a sand layer from leaking into a stone layer, while allowing gases and fluids to pass. They can also be used for reinforcement.

Hazardous waste: waste that contains compounds toxic or hazardous to humans or to the environment; that is, waste that can catch fire easily (ignitable), is corrosive to skin tissue or metals, is unstable (reactive), or contains harmful concentrations of toxic materials that can leach out.

HDPE: high-density polyethylene, a polymeric membrane in the form of sheets, used in landfill covers and bottom liners.

Hematopoietic: blood-attacking.

Hepatotoxic: liver-attacking.

Hydraulic conductivity: a measure of the ability of a substance to transmit water.

ICI: Industrial, Commercial, and Institutional; for example, ICI waste.

Impermeable: does not allow a fluid to pass through; e.g., an impermeable landfill liner.

Ionizing radiation: alpha, beta, and gamma radiation with sufficient energy to dislodge electrons from atoms, forming charged ions that can react with and damage living tissue.

Isotopes: different forms of the same chemical element which are distinguished by having different numbers of neutrons, but the same number of protons, in the nucleus of their atoms. A single element may have many isotopes.

LC_{50}: the lethal concentration of a chemical, at which 50% of exposed organisms die in a laboratory test.

LD_{50}: the lethal dose of a chemical it takes to kill 50% of the organisms exposed in a laboratory test. This parameter describes the toxicity of a substance.

Leachate: a fluid formed in a landfill.

Leachate collection system: an engineered system of pipes placed above the bottom liner of a landfill to collect and pump leachate from the landfill.

LLRW: low-level radioactive waste.

Mercaptans: substances belonging to the group of organic compounds resembling alcohols, but with the oxygen of the hydroxyl group replaced by sulfur. Many of these compounds are characterized by a strong, repulsive odor.

Methanogenic: producing methane; methanogenic bacteria create methane from organic compounds.

Microbiological waste: see Biomedical Waste.

Mil: one-thousandth of an inch.

MOE: Ministry of Environment (Ontario, Canada).

MOEE: Ministry of Environment and Energy (Ontario, Canada).

MRF: materials recovery facility, where recyclable products are separated, packaged, and stored until shipped to market.

Nephrotoxic: kidney-attacking.

Neurotoxic: nervous system-attacking.

NIMBY: Not In My Back Yard, a social phenomenon of strong local resistance to accepting landfills and other undesirable facilities in the community.

Opacity: optical density of a substance; the opposite of transparency. Smoke with high opacity is difficult to see through.

PAH: polyaromatic hydrocarbons.

Pathogen: an organism or substance that produces disease.

PCB: polychlorinated biphenyls.

Permeability: the capacity of a soil or rock for transmitting fluids.

pH: numeric measure of the relative acidity or alkalinity of a substance on a scale of 0 to 14, with the neutral point at 7. Acid solutions have pH values below 7, and basic solutions have pH values greater than 7.

Piezometer: a device placed in a borehole to measure subsurface groundwater head or pressure.

PM_{10}: particulate matter that is less than 10 microns in diameter.

Polymers: long-chain molecules made up of many small molecules.

Pozzolans: siliceous substances, such as fly ash and pumice, used in making cement.

ppm: parts per million.

Precambrian: period of geologic time before the Paleozoic, that is, older than about 570 million years.

RCRA: Resource Conservation and Recovery Act, a U.S. federal law regulating landfills.

RDF: refuse-derived fuel, waste that has been processed to form a uniform fuel for incinerators.

Reduction: a process of cooking garbage used at the turn of the century to extract a variety of marketable byproducts, such as grease.

Risk assessment: the process of gathering data and making assumptions to estimate quantitatively the harmful effects on human health or the environment from exposure to hazards associated with the use of a particular substance or technology.

Sanitary landfill: an engineered method of disposing solid wastes on land by spreading the waste in thin layers, compacting it to the smallest practical volume, and covering it with soil by the end of each working day.

Scrubber: pollution-control device used with incinerators (and fossil-fuel power plants) that removes acid gases by injecting a lime slurry into the exhaust gases.

Sustainable development: development that meets the needs of the present without compromising the ability of future generations to meet their own needs.

TDS: total dissolved solids.

Till: unsorted sediment deposited by a glacier.

Tipping fees: the charge, usually per tonne, for depositing waste at a landfill.

Transuranic isotopes: isotopes created when uranium-235 absorbs alpha or beta particles. These generally have very long decay times of thousands of years.

Tuff: volcanic ash that has been compressed so that it becomes rock.

Virus: infectious agents much smaller than bacteria which need to grow in an animal, plant, or bacterial cell. Some viral infectious diseases include smallpox, measles, chicken pox, the common cold, rabies, and viral pneumonia.

VOC: volatile organic compounds.

Windrow: organic materials placed in long rows for composting.

WTE: waste-to-energy, referring to waste incinerators that generate electricity.

REFERENCES

Anonymous. 1995. Vinyl Chloride Found in Landfill Gas. *Hazardous Materials Management* 7(3): 51.

Apogee Research. 1995. *The Canadian Hazardous Waste Inventory (CHWI)*. Prepared for Environment Canada.

Barlaz, M. A., R. K. Ham, and D. M. Schaefer. 1990. Methane Production from Municipal Refuse: A Review of Enhancement Techniques and Microbial Dynamics. *Critical Reviews in Environmental Control* 19(6): 557–584.

Barlaz, M. A., and R. K. Ham. 1993. Leachate and Gas Generation. In *Geotechnical Practice for Waste Disposal*, edited by D. Daniel. New York: Chapman and Hall.

Birmingham, B., et al. 1996. Environmental Risks of Municipal Waste Landfilling and Incineration. Presented by Ontario Ministry of Environment and Energy at MSW Incinerators: Burning Questions Seminar, Toronto, 1 May 1996.

Blumberg, L., and R. Gottlieb. 1989. *War On Waste: Can American Win Its Battle with Garbage?* Washington, D.C.: Island Press.

Boraiko, A. A. 1985. Storing Up Trouble . . . Hazardous Waste. *National Geographic* 167(3): 318–351.

Brosseau, J., and M. Heitz. 1994. Trace Gas Compound Emissions from Municipal Landfill Sanitary Sites. *Atmospheric Environment* 28(2): 285–293.

Brown, K. W., and K. C. Donnelly. 1988. An Estimation of the Risk Associated with the Organic Constituents of Hazardous and Municipal Waste Landfill Leachates. *Hazardous Waste and Hazardous Materials* 5(1): 1–30.

Carlsson, A. 1990. Nuclear Repository Under the Sea. *Tunnels and Tunnelling* (September), pp. 50–51.

Carson, Rachel. 1962. *Silent Spring.* Cambridge, Mass.: Riverside Press.

Chem-Security (Alberta), Ltd. n.d. *Alberta Special Waste Management System.* [Corporate brochures on incineration, landfilling, and treatment.]

Clark, R. B. 1989. *Marine Pollution.* Oxford: Clarendon.

Clark, G. W. 1989. Rabbit Lake Project: Mining and Development. *Canadian Institute of Mining and Metallurgy Bulletin* (December), pp. 49–58.

Costanza, R., R. d'Arge, R. deGroot, S. Farber, M. Grasso, B. Hannon, K. Limburg, S. Naeem, R.V. O'Neill, J. Paruelo, R.G. Raskin, P. Sutton, and M. van den Belt. 1997. The Value of the World's Ecosystem Services and Natural Capital. *Nature* 387(15): 253–260.

Crittenden, G. 1995. Dioxin: 1 Reality: 0—Politics vs. Science in the Canadian Media Wars. *Hazardous Materials Management* 7(4): 71–74.

Crutcher, A. J., and J. R. Yardley. 1992. Implications of Changing Refuse Quantities and Characteristics on Future Landfill Design and Operations. In *Municipal Solid Waste Management*, edited by M. E. Haight. Waterloo, Ont.: University of Waterloo Press.

Darilek, G., R. Menzel, and A. Johnson. 1995. Minimizing Geomembrane Liner Damage while Emplacing Protective Soil. In *Geosynthetic '95*, vol. 2. Nashville, Tenn.: IFA I.

Department of Energy. 1996. *Report to Congress: 1995 Annual Report on Low-Level Radioactive Waste Management Progress*. DOE/EM-0292. Washington, D.C.

Dormuth, K. W., and K. Nuttall. 1987. The Canadian Nuclear Fuel Waste Management Program. *Radioactive Waste Management and the Nuclear Fuel Cycle* 8(2–3): 93–104.

Duffy, D. 1994. Transforming a Dump: Makeover at Fresh Kills. *Conservationist* 47(3): 54–55.

Dusseault, M. B. 1995. Slurry Fracture Injection. *Hazardous Materials Management* 7(1): 16–18.

ECDC Environmental, Laidlaw Environmental Services. n.d. *The ECDC Landfill Vision*. [Corporate brochure describing the ECDC landfill, Salt Lake City.]

Eedy, W., and D. Hart. 1988. *Estimation of Long-term Probabilities for Inadvertent Intrusion into Radioactive Waste Management Areas: A Review of Methods*, INFO-0275. Ottawa: Atomic Energy Control Board.

Emcon Associates. 1980. *Methane Generation and Recovery from Landfills*. Ann Arbor, Mich.: Ann Arbor Science Publishers.

Environment. 1987. Landfills: A New Source of Global Methane. *Environment* 29(3): 21–22.

Environmental Protection Agency (EPA). 1985. *Report to Congress: Injection of Hazardous Waste*. EPA 570/0-85-003. Washington, D.C.

Environmental Protection Agency (EPA). 1994. *Characterization of Municipal Solid Waste in the United States: 1994 Update*. EPA 530-S-94-052. Washington, D.C.

Environmental Protection Agency (EPA). 1996. *Standards of Performance for New Stationary Sources, and Guidelines for Control of Existing Sources, Municipal Solid Waste Landfills*. 40 CFR, Parts 51, 52, 60. *Federal Register* 61, no. 49 (March 12). Washington, D.C.

Flosdorf, H. W., and S. Alexieff. 1993. Mining Landfills for Space and Fuel. *Solid Waste and Power Journal* (March/April), pp. 26–32.

Getz, N. P. 1994. How Does Waste-To-Energy "Stack" Up? *Journal of the Air and Waste Management Association* 44: 1309–1312.

Griffin, R. D. 1988. *Principles of Hazardous Materials Management*. Chelsea, Mich.: Lewis.

Guelph, City of. n.d. *Wet/Dry Recycling Centre—Facts.* [Fact sheet and statistics.] Guelph, Ont.

Health and Welfare Canada. 1993. *Guidelines for Canadian Drinking Water Quality.* 5th edn. Ottawa: Canada Communications Group Publishing.

Henry, J. G., and O. J. C. Runnals. 1989. Hazardous Wastes. In *Environmental Science and Engineering*, pp. 538–581, edited by J. G. Henry and G. W. Heinke. Englewood Cliffs, N.J.: Prentice-Hall.

Hershkowitz, Allen, and Eugene Salerni. 1987. *Garbage Management in Japan: Leading the Way.* New York: Inform Inc.

Johnson, L. J., et al. 1994. *The Disposal of Canada's Nuclear Fuel Waste: Engineered Barriers Alternatives.* Atomic Energy of Canada Limited Report AECL-10718, COG-93-8.

Jones, Kay H. 1994. Comparing Air Emissions from Landfills and WTE Plants. *Solid Waste Technologies* (March/April), pp. 28–39.

Journal of Waste Recycling. 1991. *The Biocycle Guide to the Art and Science of Composting.* Emmaus, Pa.: JG Press.

Koerner, R. M. 1994. *Designing with Geosynthetics.* 3rd edn. Englewood Cliffs, N.J.: Prentice-Hall.

Lafferty, Vera. 1996. Reflections on the Process Used to Site a Low-Level Radioactive Waste Disposal Facility in Deep River, Ontario, Canada. Presented to the Appalachian Compact Users of Radioactive Isotopes (ACURI) Seventh Annual Meeting, Harrisburg, Pennsylvania, August 1996.

Lahl, U., et al. 1990. PCDD/PCDF Balance of Different Municipal Waste Management Methods, Five Parts: I, Municipal Waste Incinerators; II, Waste Disposal and Disposal Gas Incineration; III, Composting; IV, Recycling; V, Comparison and Results. Presented at Dioxin 90, Bayreuth, Germany, August 1990.

Lancaster County Solid Waste Management Authority. n.d. *Resource Recovery Facility*]Brochure, Lancaster, Pennsylvania.]

Lee, G. F., and R. A. Jones. 1991. Landfills and Ground-water Quality. *Groundwater* 29(4): 482–486.

Loder, T. C., F. E. Anderson, and T. C. Shevenell. 1983. *Sea Monitoring of Emplaced Baled Solid Waste.* University of New Hampshire Report S.D.-118.

Lowry, D. and K. C. Chan. 1994. New Approach to Waste Management Prevents Leachate Migration. *Environmental Science and Engineering* (June/July), p. 16.

Malone, D., ed. 1936. George E. Waring. In *Dictionary of American Biography*, vol. 10. New York: Charles Scribner's Sons.

Management Information Services, Inc. 1994. *Economic and Employment Benefits of the Use of Radioactive Materials: The Untold Story.* Prepared for Organizations United for Responsible Low-Level Radioactive Waste Solutions. Washington, D.C.

McBean, E. A., F. A. Rovers, and G. J. Farquhar. 1995. *Solid Waste Landfill Engineering and Design.* Englewood Cliffs, N.J.: Prentice-Hall.

Miller, G. T., Jr. 1997. *Living in the Environment.* 7th edn. California: Wadsworth.

Mine Reclamation Corporation. 1995. *Eagle Mountain Landfill and Recycling Centre: Project Description.* Revised April 14, 1995. Palm Desert, Calif.

Ministry of Environment (MOE). 1991a. *Waste Disposal Site Inventory.* PIBS 256. Toronto.

Ministry of Environment (MOE). 1991b. *Interim Guidelines for the Production and Use of Aerobic Compost in Ontario.* PIBS 1749. Toronto.

Ministry of Environment and Energy (MOEE). 1993. *Guidance Manual for Landfill Sites Receiving Municipal Wastes.* PIBS 2741. Toronto.

Murphy, Pamela. 1993. *The Garbage Primer: A Handbook for Citizens.* New York: League of Women Voters/Lyons and Burford.

Neal, H. A., and J. R. Schubel. 1987. *Solid Waste Management and the Environment: The Mounting Garbage and Trash Crisis.* Englewood Cliffs, N.J.: Prentice-Hall.

Newberry, W. F. 1993. The Rise and Fall and Rise and Fall of American Public Policy on Disposal of Low-Level Radioactive Waste. *South Carolina Environmental Law Journal* (Winter), pp. 57–87.

New York City Department of Sanitation. 1994a. *Fresh Kills Landfill Leachate Mitigation/Corrective Measures Assessment.* DOS Fact Sheet vol. 1, no. 6 (March).

New York City Department of Sanitation. 1994b. *Stability Investigation and Monitoring at Fresh Kills Landfill.* DOS Fact Sheet, vol. 1, no. 2 (February).

New York City Department of Sanitation. 1994c. *Landfill Gas Management.* DOS Fact Sheet, vol. 1, no. 4 (February).

Nuclear Energy Insitute. 1996. *Survey of Expenditures for Implementation of the Low-Level Radioactive Waste Policy Act as Amended.* Washington, D.C.

O'Leary, P. R., L.Canter, and W. D. Robinson. 1986. Land Disposal. In *The Solid Waste Handbook: A Practical Guide,* edited by W. D. Robinson, pp. 259–376. New York: John Wiley and Sons.

Organizations United for Responsible Low-Level Radioactive Waste Solutions. 1996. *Lessons Learned from the Barnwell Closure to 31 States.* Washington, D.C.

Oweis, I. S., and R. P. Khera. 1998. *Geotechnology of Waste Management.* Boston: PWS Publishing.

Parmeggiani, L., ed. 1983. *Encyclopaedia of Occupational Health and Safety.* 3rd edn. Geneva: International Labour Office.

Platt, B., C. Doherty, A. C. Broughton, and D. Morris. 1991. *Beyond 40%: Record-Setting Recycling and Composting Programs.* Washington D.C.: Institute for Local Self-Reliance/Island Press.

Pohland, F. 1989. Properly Designed Landfills May be Safe Disposal Sites for Some Hazardous Wastes. *American Journal of Public Health* 79(1): 215.

Pratt, M. G. 1995. Energy Recovery vs. Landfilling. *Hazardous Materials Managment* 7(1): 29–33.

Rabe, B. G. 1992. Beyond the NIMBY Syndrome in Hazardous Waste Facility Siting: The Albertan Breakthrough and the Prospects for Cooperation in the United States and Canada. In *Tensions at the Border: Energy and Environmental Concerns in Canada and the United States,* edited by J. Lemco, pp. 141–163. New York: Praeger.

Rabe, B. G. 1994a. Siting Success: Replication of the Initial Canadian Breakthrough in Hazardous Waste Facility Siting. *Journal of Resource Management and Technology* 22(1): 1–9.

Rabe, B. G. 1994b. *Beyond NIMBY: Hazardous Waste Siting in Canada and the United States.* Washington, D.C.: Brookings Institute.

Rathje, W. L., and L. Psihoyos. 1991. Once and Future Landfills. *National Geographic* 179(5): 116–134.

Rathje, W. L., and C. Murphy. *Rubbish! The Archaeology of Garbage.* New York: Harper Collins.

Reynolds, A. B. 1996. *Bluebells and Nuclear Energy.* Madison, Wis.: Cogito Books.

Royal Commission on Environmental Pollution. 1993. *17th Report. Incineration of Waste.* London: H.M. Stationary Office.

Rylander, H. 1994. *Integrated MSWM Systems around the World: Sweden.* Silver Spring, Md.: Solid Waste Association of North America.

Sarofim, A. F. 1997. Thermal Processes: Incineration and Pyrolysis. In *Handbook of Solid Waste Management*, edited by D. G. Wilson, pp. 166–196. New York: Van Nostrand Reinhold.

Schroeder P. R., C. M. Loyd, P. A. Zappi, and N. A. Aziz. 1994. *The Hydrologic Evaluation of Landfill Performance (HELP) Model, Version 3.* EPA/600/R-94/168a. Cincinnati, Ohio: U.S. Environmental Protection Agency.

Sirman, I. A. 1995. Canadian Leachate Treatment Options. *Hazardous Materials Management* 7(4): 49–52.

SKB. Swedish Nuclear Fuel and Waste Management Company. 1992. *Mechanical Integrity of Canisters.* Report SKB-TR-92-45. Stockholm.

SKB. Swedish Nuclear Fuel and Waste Management Company. n.d. *Swedish Final Repository for Radioactive Waste—SFR.* [Corporate brochure, Stockholm.]

Stone, R. 1977. Sanitary Landfill. In *Handbook of Solid Waste Management*, edited by D. G. Wilson, pp. 226–263. New York: Van Nostrand Reinhold.

Tammemagi, H. Y., and S. N. Thompson. 1990. International Perspectives on Low-Level Radioactive Waste Disposal. In *Proceedings, Waste Management '90*, vol. 1, pp. 589–594, edited by Roy G. Post. Tucson, Arizona.

Taplin, D., and F. B. Claridge. 1987. *Performance of Engineered Barriers for Low-Level Waste.* Prepared by Geotechnical Resources Ltd. and Komex Consultants Ltd. for the Atomic Energy Control Board. INFO-0274. Ottawa.

Tchobanoglous, G., H. Thiesen, and S. Vigil. 1993. *Integrated Solid Waste Management: Engineering Principles and Management Issues.* New York: McGraw-Hill.

Turner, A. Personal Communication, Marketing Manager, Chem-Security (Alberta) Ltd., Calgary, 17 August 1998.

U.S. Forest Service, U.S. Department of Agriculture. 1984. *Pesticide Background Statements*, vol. 1, *Herbicides.* Agriculture Handbook No. 633, Washington, D.C.

Virtanen, Y., and S. Nilsson. 1993. *Environmental Impacts of Waste Paper Recycling.* London: International Institute for Applied Systems Analysis/Earthscan Publications.

Walsh, D. C. 1991. The History of Waste Landfilling in New York City. *Groundwater* 29(4): 591–593.

Whitaker, J. S. 1994. *Salvaging the Land of Plenty: Garbage and the American Dream.* New York: William Morrow.

White, B. 1990. Status of Landfill Methane Recovery Projects. *Gas Energy Review* 18(7): 2–5.

Wilson, D. G. 1977. History of Solid Waste Management. In *Handbook of Solid Waste Management*, edited by D. G. Wilson, pp. 1–9. New York: Van Nostrand Reinhold.

Woods, R. 1991. Ashes to . . . Ashes? *Waste Age* (November), pp. 46–52.

World Resources Institute, United Nations Environment Programme, and United Nations Development Programme. 1992. *World Resources 1992–93: A Guide to the Global Environment.* New York: Oxford University Press.

Yach, R. 1996. Draining Landfills: Trenchless Today. *Canadian Environmental Protection* (September).

INDEX